Wireless Power Handbook

Wireless Power Handbook

A Supplement to GaN Transistors for Efficient Power Conversion

Second Edition

Michael A. de Rooij, Ph.D.

Power Conversion Publications
El Segundo MMXV

ISBN: 978-0-9966492-1-6

Library of Congress Control Number: 2015953973

Second Edition

I dedicate this book:

To my spouse Nicholas who showed incredible patience over the many hours it took to complete this work.

To my parents Jacob and Emma who supported me with my education.

To my many colleagues who helped me gain the requisite knowledge and experience to understand the subject.

To Alex Lidow and EPC for this opportunity.

And lastly, to my late four-legged friend, Caramelle, who kept my feet warm under the table.

Table of Contents

Preface to the Second Edition

This second edition comes less than a year following the release of the first edition – such is the pace at which the understanding and application of wireless power transfer is moving. It is truly a "high tech" force that is being pulled by the market demand. "Cut the cord" is their battle cry...and now that they know it can be done, ask what's holding us back, let's pick up speed and get on with it!

Ongoing efforts to improve the performance of wireless power have led to the development of specifically designed devices and new control techniques. One result of this development is a fully A4WP-compliant class 2 demonstration system that does not need adaptive matching. This amounts to huge cost savings for the system through circuit complexity reduction. Ongoing system level tests also revealed flaws in the wireless power standards that require correction; all in an effort to improve the wireless power experience.

The most difficult hurdle faced by wireless power system designers has been the abatement of radiated EMI. Given that the wireless power system is an intentional radiator and many power electronic designers struggle with this issue, this second edition therefore goes into greater detail on the filtering requirements for radiated EMI reduction.

Support for multi-mode wireless systems was also expanded to improve implementation and amplifier design support, as a single amplifier solution is required to keep adoption of wireless power moving forward. A new topic was added that covers control strategies for wireless power systems. This new topic was added in an effort to stimulate IC manufacturers to increase their efforts to support A4WP products and show product designers potential solutions using standard techniques.

While the scope of this second edition has expanded, there is still much more to accomplish, as this fast moving technology evolves I sign off to prepare for what comes next. Thank you for your interest in this work.

Michael A. de Rooij
El Segundo, California
October 2015

Preface to the First Edition

Cutting the power cord, the dawn of wireless power is here!

Since Nikola Tesla first experimented with wireless power during the early years of the 20th century, there has been a quest to "cut the cord" of electrical power – and go wireless! Over 100 years later, we finally have the technological capability to achieve Telsa's vision. Highly-resonant wireless power transfer, based on the generation of magnetic fields, has proven to be a viable path. Magnetic fields offer the necessary requisites – ease of use, robustness and, most importantly it is considered safe.

Applications for wireless power are endless, from mobile communications and computing, powering systems in hazardous environments to implantable medical devices. Wireless power systems have the ability to dramatically improve the quality of life for patients needing powered implants. Imagine not having invasive wires penetrating through the skin to power artificial heart pumps.

With the explosion in the variety and number of mobile devices, wireless power transfer offers convenience of charging batteries without the annoyance of cumbersome cables and the inconvenience of "plugging in." The strong push for convenience and power on-demand by the consumer is driving the desire for this technology.

The technology may have caught up with the concept of wireless power, but the implementation poses many challenges to power system designers, even though among engineers the transmission and reception of magnetic fields are well understood. At the heart of highly-resonant wireless power is the amplifier, which drives the coils that generate the magnetic field.

Thus one of the major challenges for implementing wireless power is the design of the amplifier. What is the best amplifier topology to safely and efficiently use in a wireless power system? What can be done about other design issues, such as the effects of varying load conditions and complying with various standards such as radiated EMI regulations? Understanding these many challenges for the most optimal amplifier design that can be used in a wireless power transfer system is the aim of this handbook.

Over the past several years, three standards have emerged. These standards, being put forth by industry consortia include the Wireless Power Consortium's Qi, the Power Matters Alliance and the Alliance for Wireless Power (A4WP), which is also known as Rezence®. The Rezence standard centers on the use of a highly-resonant approach that allows loosely-coupled transmission and reception between the source and the receiving coils. This standard was chosen to drive the selection criteria for several amplifier topologies (class E, traditional voltage-mode class D, current-mode class D, and ZVS class D) that were designed, built, tested and compared for this handbook.

The A4WP standard operates in the open industrial, scientific and medical (ISM) frequency band at 6.78 MHz, which allows unlimited power in most localities. Many aspects of the design are covered in this book; from coil tuning, and the impact of load impedance variations on the amplifier, to automated retuning of the coil to keep the system efficiency high.

From those experimental results, it was clear that the ZVS class D topology, fitted with eGaN FETs, exhibited superior performance as compared to the other amplifiers. Advanced characteristics and methodologies such as radiated EMI, multi-mode systems and ways to improve efficiency are also covered.

As a handbook, it is meant to be read as a guide to designing an efficient amplifier for a wireless power transfer system. It shares experiences that could save time and point to references important to understanding and designing wireless power systems.

I want to thank my colleagues at Efficient Power Conversion Corporation and all the designers of power systems that have gone before me, confronting the design challenges and paving the way for wireless power to become a reality.

Finally…cut the power cord, wireless power is here!

CHAPTER 1:
Overview of Wireless Power Transfer

What is Wireless Power Transfer?

The transfer of electrical energy without using conductors as the transport medium

Examples of wireless power media:
- Electric fields
- Photons
- Magnetic fields

What is Wireless Power Transfer?

Energy can be transferred in many ways to yield a desired result. Cooking food is a good example of wireless power transfer in which an electric stove is used to heat food, but it is not very useful when trying to power portable electronic devices. To provide useful electrical energy for portable devices, without using electrical conductors, several media such as photons, electric fields, or magnetic fields can be used.

Each of these media has advantages and disadvantages. Photons are not efficient at transferring energy where the best-in-class may ultimately achieve only 40% – 50% when lasers are used [1.1], but typically fall around 10% [1.2]. Electric fields can be transmitted over a good distance, but cannot penetrate some materials, limiting their use. Magnetic fields offer a safer alternative, but yield shorter distances between the source and device due to their inherent loop characteristic (This characteristic is mathematically referred to as curl).

Considering the scenario where wireless power transfer is desired, most devices that need to receive power will be placed on a wireless power surface, which means there is some control over the distance between the source and the devices. And, the distance will not be excessive. Based on these comparative factors – electrical transfer, safety, and distance – the logical choice of a wireless power medium is the magnetic field.

[1.1] J-G. Werthen and M. Cohen, "The Power of Light: Photonic Power Innovations in Medical, Energy and Wireless Applications," *Photonics Spectra Magazine*, May 2006.

[1.2] A. Polman and H. A. Atwater, "Photonic design principles for ultrahigh-efficiency photovoltaics," *Nature Materials*, Vol. 11, March 2012, pp. 174–177.

Why Wireless Energy?

Mobile device charging
- Convenience of use
- Extended usable battery life

Medical implants
- Quality of life improvement
- Reduces risk of infection

Hazardous environments
- Explosive atmosphere
- Corrosive locations
- High voltage

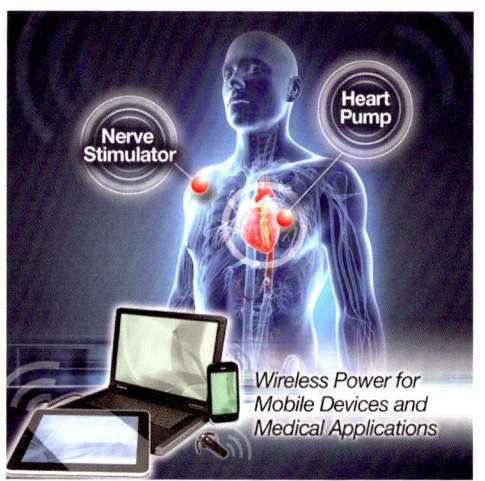

Wireless Power for Mobile Devices and Medical Applications

Why Wireless Energy?

With the explosion of the variety and number of mobile devices, wireless power transfer offers convenience of charging batteries without the annoyance of cumbersome cables, and the inconvenience of "plugging in." Additionally, wireless power could potentially extend the working life of the battery by providing untethered power on demand.

Another end-use of wireless power transfer can be found in medical applications, particularly medical implants. These rapidly emerging applications can result in major quality of life improvements and have significant life-extending implications. Imagine not having invasive wires penetrating the skin to power artificial heart pumps, but rather being able to power the pump from a remote energy source as you sit in a chair, walk around, or lie in bed.

Wireless power transfer can also be used in safety-critical environments such as explosive or corrosive atmospheres (an electrical spark in the vicinity of a gas pump comes to mind), underwater, or any location where there is a safety risk when an electrical connection is made or broken with a corresponding spark.

Characteristics of a Magnetic Field

Magnetic fields:

+ Considered safe

+ Well understood – easy to generate and capture

− Have limited efficient transmission distance – depends upon transmitter and receiver diameters

Characteristics of a Magnetic Field

Having justified the practical need for wireless power transfer and the use of magnetic fields as the transfer medium, next we need to understand the relevant characteristics of magnetic fields. First, and most importantly, magnetic fields are considered safe for use even at the frequencies targeted for wireless power transfer [1.3]. Specific absorption rate (SAR) guidelines provide the required field density limits to ensure human safety when exposed to magnetic fields and are governed by well-researched standards [1.4]–[1.6].

Secondly, among electrical engineers, magnetic fields are well understood, making them easy to generate and capture.

Lastly, magnetic fields do not transfer energy well over long distances, which is primarily due to their divergent characteristics over distance. This makes it difficult to capture enough magnetic flux the further from the source the receiver is placed. This limitation is not severe, given that most wireless power transfer applications require relatively short distances (e.g., less than 18 inches).

[1.3] J. Nadakuduti, L. Lu, P. Guckian, "Operating Frequency Selection for Loosely Coupled Wireless Power Transfer Systems with Respect to RF Emissions and RF Exposure Requirements," *IEEE Wireless Power Transfer Conference*, May 15–16, 2013 Perugia, Italy, pp 234–237.

[1.4] *Class B–Human Exposure Limits*, FCC Part 1.1310.

[1.5] *Human Exposure Limits* – Recommendation 1999/519/EC.

[1.6] *Human Exposure Limits* – ICNIRP 2010.

Challenges to Wireless Power Transfer

High efficiency – limited power dissipation budget

Low profile – needed for the mobile market

Robust to dynamic operating conditions

Defined response to foreign metal objects

Compliance to regulatory standards

Challenges to Wireless Power Transfer

The implementation of a wireless power transfer system poses many challenges to power system designers. Some of the challenges are market-driven, while others are related to the practicality of the system. Today the mobile gadget market is driving the development of wireless power transfer, thus setting many of its requirements and challenges.

These requirements include high efficiency, particularly for the receiving devices due to limited power dissipation budgets, low physical profile, and robustness to all operating conditions. The need for robustness stems from the convenience-of-use factor that wireless power transfer offers – users do not want to be burdened with rules on device placement, limitations on the number of devices that can be powered at one time, and the size of the devices to be powered. Add to these requirements the need for systems to anticipate adverse operating conditions, such as the introduction of foreign objects that can drastically affect the operation and performance of wireless power transfer systems.

Lastly, these systems need to conform to EMI emission standards such as FCC part 18 [1.7], and the equivalent EN standards such as EN 55011 [1.8] and EMC directive (2004/108/EC) [1.9].

[1.7] *FCC Code of Federal Regulations Title 47*, Vol. 1, Part 18 B (Industrial, Scientific, and Medical Equipment), 1998.

[1.8] European Norm. EN55011 Group 2 Class B.

[1.9] *Electromagnetic Compatibility (EMC)*, European Directive (2004/108/EC).

Wireless Power Transfer Standards Overview

| Standard | rezence
Alliance For Wireless Power | *pma
Power Matters Alliance™ | Qi WIRELESS POWER
CONSORTIUM |
|---|---|---|---|
| Frequency | 6.78 MHz | ~201 – 315 kHz | ~ 100 – 205 kHz |
| Power | 1 W – 70 W | 5 W | 5 W, 10 W, 15 W |
| Placement | Any orientation, any placement | Specific placement | Specific placement |
| Multiple devices | Yes | No | Limited |
| Operating Principle | Resonance | Inductive | Inductive, some resonance |
| Communications | Bluetooth Low Energy | In-Band | In-Band |

*Merged with rezence

Wireless Power Transfer Standards Overview

Most of the older wireless power solutions focused on tight coupling, with induction coil solutions operating at relatively low frequencies from 100 kHz through 315 kHz. This is the basis of the Qi (Wireless Power Consortium) and Power Matters Alliance (PMA) standards.

The Alliance for Wireless Power (A4WP) standard, called Rezence [1.10], makes use of high-frequency (6.78 MHz) operation that allows resonance to be used to enhance the generation and transmission of magnetic fields for wireless power transmission [1.11, 1.12]. This use of high-frequency operation is the basis for the loosely-coupled, highly-resonant approach to wireless power transfer. There are many advantages to this approach that will become apparent as we delve more into the subject. The two most important advantages are supporting multiple devices simultaneously, including various power levels, and the placement orientation of the devices. In contrast, the Qi standard now supports devices of various power levels, but only one device at a time.

The Rezence standard has targeted a wide range of power levels to address the multitude of device sizes. The Qi standard has also addressed higher power to some degree and now can support up to 15 W. This was achieved since the Qi standard also makes limited use of resonance to enhance performance.

In all formats, power management and control between the source and device (that is, transmitter and receiver) is established using digital communications. In the case of the Wireless Power Consortium (WPC) Qi standard and the Power Matters Alliance (PMA) standards, the digital information is encoded on the power carrier. Whereas, in the Rezence standard, use is made of the Bluetooth Low Energy (BLE) standard, as the ISM band restriction of 15 kHz bandwidth precludes power modulation as a viable communications option. The BLE solution makes the Rezence standard a more universal solution than the Qi and PMA standards.

[1.10] R. Tseng, B. von Novak, S. Shevde and K. A. Grajski, "Introduction to the Alliance for Wireless Power Loosely-Coupled Wireless Power Transfer System Specification Version 1.0," *IEEE Wireless Power Transfer Conference 2013, Technologies, Systems and Applications*, May 15–16, 2013.

[1.11] A. Karalis, J.D. Joannopoulos, M. Soljačić, "Efficient wireless non-radiative mid-range energy transfer," *Annals of Physics*, Vol. 323, No. 1, 2008, pp. 34–48.

[1.12] A. Kurs, A. Karalis, R. Moffatt, J. D. Joannopoulos, P. Fisher, M. Soljačić, "Wireless Power Transfer via Strongly Coupled Magnetic Resonances," *Science*, Vol. 317 No. 6, July 2007, pp. 83–86.

Criteria for the Selection of a Wireless Power Transfer Standard

What markets, less than 50 W, can the standard target?
- Mobile communications
- Computing
- Low-power medical

Does the standard address the "convenience factor" for the user?
- Only A4WP standard addresses this factor

Criteria for the Selection of a Wireless Power Transfer Standard

The choice of a wireless standard for the design of a specific power transfer system needs to consider many factors, with the power level and target applications being typically the two dominant factors. In the case of medical and mobile computing applications, key additional factors are safety and convenience of use.

The Qi and PMA standards have drawbacks, such as the need for precise placement of the device on the source, as well as the ability of the source to drive only one device at a time. Whereas, the Rezence standard uses magnetic resonance which makes it possible to have a single source capable of delivering power to multiple devices simultaneously, regardless of the orientation of the receiving devices. In addition, using resonance allows the system to deliver higher power than the inductive-based standards.

Wireless Power Transfer Selection

A4WP (Rezence®) was selected as the standard to be used because it is:

- Highly resonant – improves transmission of energy

- Allows loose coupling between source and device – addresses the convenience factor for the user

- Operates using unlicensed ISM band frequency of 6.78 MHz

Wireless Power Transfer Selection

Having compared various wireless power standards, the decision was made to adopt the A4WP Rezence standard as the primary subject of this handbook. This standard is characterized by being highly resonant, allowing loose coupling between the source and the device. Further, the A4WP standard operates in the open industrial, scientific and medical (ISM) frequency band at 6.78 MHz [1.13]. This allows end users to have access to a wireless power product without a subscription for the use of the frequency band. Operation at this frequency will require careful selection of an amplifier and deliberate consideration for other design choices to ensure high efficiency. Evaluating these wireless power transfer systems' design issues is the focus of this work.

[1.13] "ISM band." *Wikipedia: The Free Encyclopedia*. Wikimedia Foundation, Inc. January 2014. [Online] Available: http://en.wikipedia.org/wiki/ISM_band

CHAPTER 2:
Wireless Power System Overview

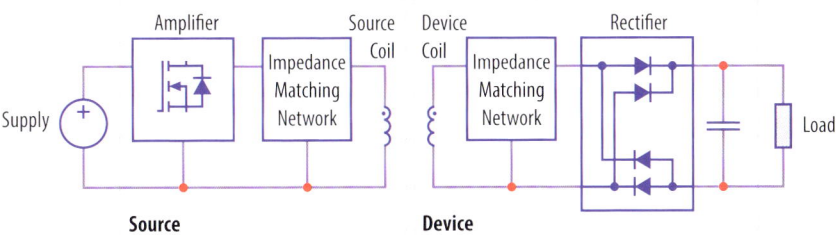

Four main sections of a wireless power transfer system:
Amplifier, Source Coil, Device Coil, and Rectifier

Wireless Power System Overview

The general architecture of a wireless power transfer system is comprised of four basic building blocks: 1) an amplifier, also known as a power converter; 2) a source coil that includes a tuning network; 3) a device coil that includes a tuning network; and, 4) a rectifier with high frequency filtering. Each of these four basic building blocks must be designed to work with adjacent blocks, which, as operating conditions vary, can prove to be challenging. Therefore, each of the challenges will be addressed and various solutions proposed. For simplicity, we will refer to the amplifier through source coil as "the source" and the device coil through output terminals as "a device."

The advantages and disadvantages of suitable amplifiers will be presented. Likewise, the impact of the choice of source and device coils on the choice of amplifier for specific power delivery will be presented, and a discussion on the impact of the convenience-of-use on the overall system will be undertaken.

The operation of the amplifier will need to adjust as power demand shifts. Also, since load variations to the amplifier are substantially larger than in traditional radio frequency (RF) wireless communication systems, more aggressive techniques are required to ensure efficiency remains within practical limits for both the amplifier and the balance of the system.

Wireless Coil-Set Overview

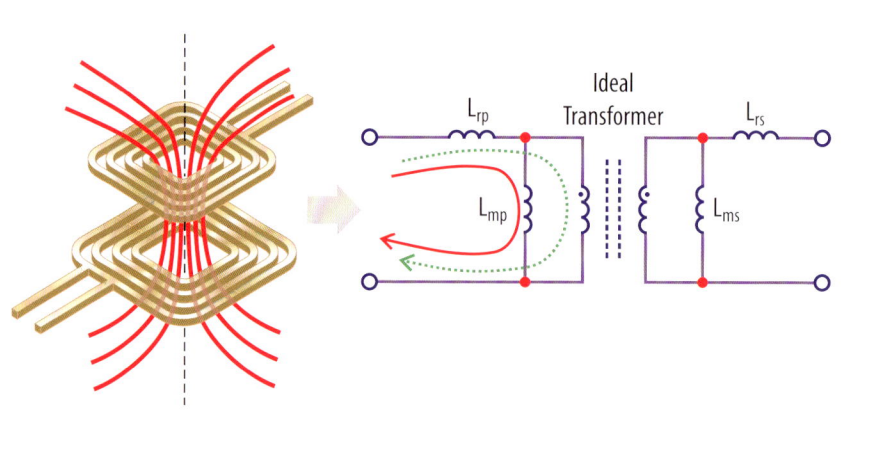

Wireless Coil-Set Overview

The basis for wireless power transfer is centered on magnetic field coils, which is shown on the left in the diagram. A set of coils forms an air core transformer whose leakage (L_{rx}) is high with respect to the magnetizing inductance (L_{mx}), as shown on the right.

Traditional transformer design ensures that the leakage component is kept as low as possible, which keeps the transmission efficiency high. In the case of wireless power transfer, the leakage component is significantly larger than the magnetizing component. Thus, use of resonance is made to increase the current into the magnetizing inductance, thereby restoring transmission efficiency.

Analysis of this circuit reveals that the ability to efficiently transfer power to the secondary side is almost entirely determined by the primary-side leakage inductance [2.1].

[2.1] K. Siddabattula, "*Wireless Power System Design Component and Magnetics Selection*," Texas Instruments. [Online] Available: http://e2e.ti.com/support/power_management/wireless_power/m/mediagallery/526153.aspx

Highly-Resonant Wireless Power Transfer

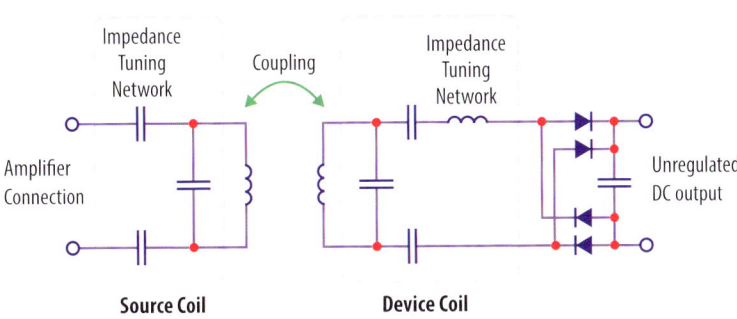

Highly-Resonant Wireless Power Transfer

To realize highly-resonant wireless transfer and to overcome the leakage inductance of the coil set, the coils are tuned to resonate at an operating frequency of 6.78 MHz, the lowest ISM band. This resonance can be achieved by using both series and parallel tuning. Series tuning increases the current (lowers the impedance) and parallel tuning increases the voltage (increases the impedance).

A balance between voltage and current requirements can be found that results in a specific impedance under specific load conditions. The choice of specific impedance should be made to yield the highest energy or some similar metric. Coupling and load variations, unfortunately, shift the tuned resonance of the coils. This shift must be taken into account when designing the amplifier and rectifier.

In the case of the device coil, an additional inductance can be added to help absorb large dynamic changes in load current. Changes in load current can result in unwanted (out of ISM band) harmonic currents being injected into the coil. These, in turn, can be radiated and cause the system to fail radiated EMI compliance testing.

In the case of the source coil, series-only tuning is almost always adopted as it yields the highest current, which is used as a control mechanism for wireless power transfer systems.

Source Coil Tuning

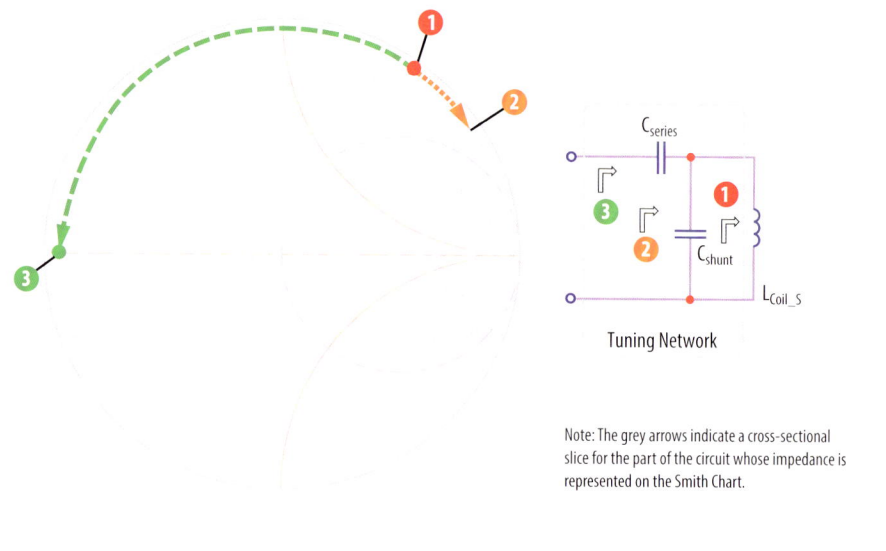

Tuning Network

Note: The grey arrows indicate a cross-sectional slice for the part of the circuit whose impedance is represented on the Smith Chart.

Source Coil Tuning

The source coil is essentially a large inductor, whose impedance is indicated by the red dot with a number "1" on the Smith Chart [2.2] at 6.78 MHz, and can be measured using a Vector Network Analyzer (VNA). If required by the connection between the source coil and amplifier, this measurement should include the transmission line.

To tune the source coil, first select a base condition, typically without a device or foreign metal object being present. This is necessary to avoid large unforeseen impedance shifts. Adding a shunt capacitor to the coil will increase the impedance (not typically employed), as shown by the orange arc arrow. Adding series capacitance to the coil will reduce its impedance, as shown by the green arc arrow. The choice of capacitance is critical because the end result must be a pure resistance shown by the green dot, which represents a tuned source coil.

The effect of coupling factor on the reflected impedance can be determined after the source and device coils are tuned and placed in proximity to each other.

[2.2] Smith Chart background image courtesy of RF Café , [Online] Available: http://www.rfcafe.com/business/software/smith-chart-for-excel/smith-chart-for-excel.htm

Device Coil Tuning

Note: The grey arrows indicate a cross-sectional slice for the part of the circuit whose impedance is represented on the Smith Chart.

Device Coil Tuning

The device coil is also essentially a large inductor, but is typically smaller than the source coil. Its impedance is indicated by the red dot on the Smith Chart at 6.78 MHz and can be measured using a VNA.

To tune the device coil, first select a base condition, typically done in the application with enclosure and other major components present. This is necessary to avoid impedance shifts that can yield significant deterioration in performance by detuning the device coil when it is installed into an application. In addition, the device coil should be tuned in the absence of a tuned source coil, to prevent tuning to a beat frequency shift as a result of the coupling between the two coils.

Adding a shunt capacitor to the coil will increase the impedance shown by the blue arc arrow. This is used to tweak the output voltage that will be supplied to the rectifier beyond the turns ratio of the source and device coils. Adding series capacitance to the coil will reduce its impedance, as shown by the green arc arrow.

Finally, the addition of series inductance results in a pure resistance shown by the orange dot with a "4" in the center on the Smith Chart, which represents a tuned device coil. The function of the inductor is to help reduce large transient load changes from inducing high out-of-band harmonic currents into the device coil that can radiate as EMI.

Transformer Model – Single Device

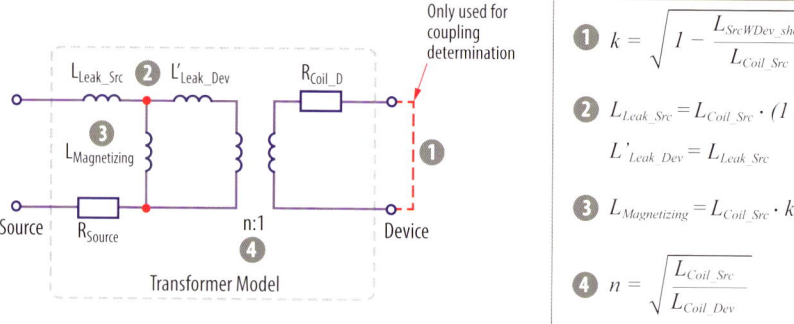

Transformer Model

$$k = \sqrt{1 - \frac{L_{SrcWDev\_short}}{L_{Coil\_Src}}}$$

$$L_{Leak\_Src} = L_{Coil\_Src} \cdot (1 - k)$$

$$L'_{Leak\_Dev} = L_{Leak\_Src}$$

$$L_{Magnetizing} = L_{Coil\_Src} \cdot k$$

$$n = \sqrt{\frac{L_{Coil\_Src}}{L_{Coil\_Dev}}}$$

Transformer Model – Single Device

A single device coil-set can be modelled as a transformer [2.3]. The components of the transformer include leakage inductances, magnetizing inductance, an ideal transformer, and winding resistances.

The model is derived from a few key measurements that can be made using a VNA. These measurements are:

1. The source coil impedance (without the device coil present) – (L_{Coil_Src})

2. The device coil impedance (without the source coil present) – (L_{Coil_Dev})

3. The source coil with device present (per application setup) and with the device in short circuit – ($L_{SrcWDev\_short}$)

Since the coupling between the coils is low, it is important to ensure a good, clean, and accurate measurement.

Transformer Model – Single Device - *continued*

The coupling coefficient can then be calculated using equation 1. This coupling coefficient can be used to determine the leakage components. It can be shown that the device side leakage (L_{Leak_Dev}) can be transformed to the source side (L'_{Leak_Dev}), and is equal to the value derived for the source side (L_{Leak_Src}). This simplifies calculations using the model. The magnetizing inductance ($L_{Magnetizing}$) can be calculated using equation 3. Finally, the turns-ratio for the model can be determined using equation 4. This is used to transform device side impedances to the source side. It is important to note that, unlike the case of classical transformer models that only result in integer turn-ratios, the coil-set transformer model can have any rational number as a result. This is due to the difference in area in the center of the coils.

The loss components for the coils are modelled using resistors that represent the winding resistance. It is important to use the resistance value at the fundamental operating frequency, which in the case of wireless power conforming to Rezence standards will be 6.78 MHz.

It is not necessary to model capacitances in this scenario, as the values between the source and device are very low and have negligible impact on the accuracy of the model.

[2.3] R. J. Smith, *Circuits, Devices and Systems*. Fourth Edition, Wiley.

Coil-Set Transformer Model – Two Devices

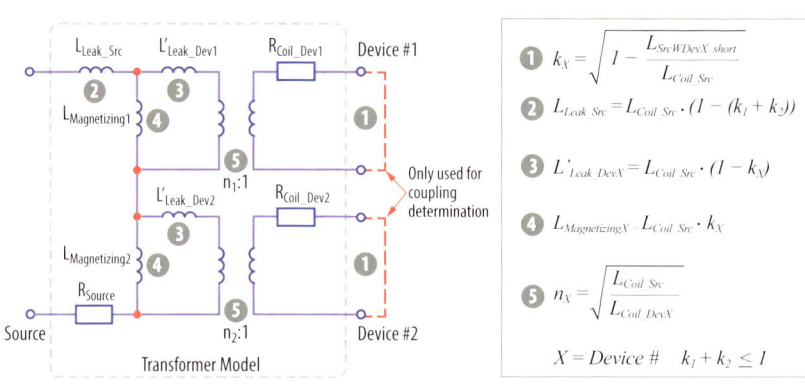

Coil-Set Transformer Model – Two Devices

A coil-set comprised of two devices can also be modelled as a transformer in a similar manner as for a single device. The devices share the magnetic flux field resulting in a series-connection of the device models. It is important to note that in this model the sum of each device's coupling coefficients cannot exceed one. The coupling coefficients may be vastly different and can represent a large device and a small device. If the coupling coefficient of one device becomes zero (the device is no longer present), then the model reduces to that of a single device case. This model ignores coupling between the devices, as it is assumed to be significantly smaller than between a device and the source. This will be true when the two devices are not placed close to each other. If the devices are large and placed close to each other, they can couple. This coupling can further be modelled in the same manner as between a source and a single device. The derivation of this model falls outside the scope of this work.

The measurement and calculations for the two device model follows the same procedure as in the case of a single device except there will be additional measurements. Only one device should be measured at a time. If coupling between devices is suspected of being significant, then additional measurements can be taken to account for the three cases; device 1 in short, device 2 in short, and both device 1 and device 2 in short. If the resulting coupling coefficient of the two devices shorted case is vastly different than the sum of the single devices in short case, then the coupling between the devices is significant and needs to be taken into account during subsequent analysis.

Device Loading

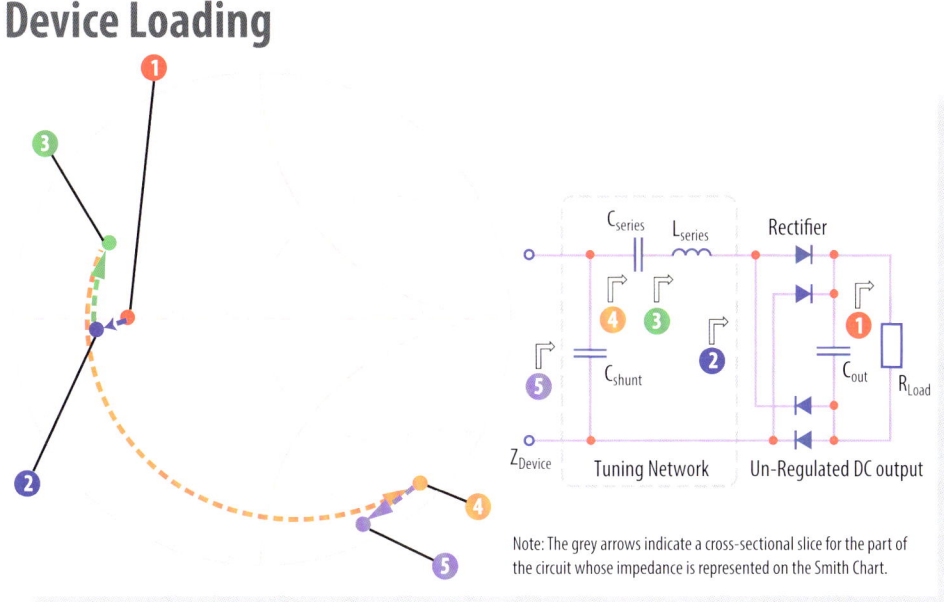

Note: The grey arrows indicate a cross-sectional slice for the part of the circuit whose impedance is represented on the Smith Chart.

Device Loading

Using the transformer model and tuning circuits, the impact of the DC load resistance (R_{Load}) variation on the tuned source coil can be analyzed. This will be done in stages starting with the device tuning circuit, and working back through the transformer model, and ending with the source tuning circuit. The end of each step will be the beginning of the next step.

A fixed value for the DC load resistance (R_{Load}) shown by the red dot (1) is used to start. This DC resistance needs to be transformed into an equivalent AC resistance, as seen looking to (2) by the blue dot. If sinusoidal voltages and currents are assumed, then the DC load resistance can simply be multiplied by $1/\sqrt{2}$ to reveal the AC load resistance. However, in practical circuits, with 3rd harmonic current crest factors introduced by the diodes and the voltage drop of the diodes, this approximation will need to be adjusted. It is best to conduct a few experiments to determine a good scaling factor to use. In addition, the capacitances of the diodes need to be considered for the equivalent AC resistance, as they appear in parallel with the input terminals of the rectifier.

Next, the impact of the two series components is determined and shown for the inductor (L_{series}) with the green dot (3) and the capacitor (C_{series}) by the orange dot (4).

The last component to be considered is the parallel capacitor ($C_{parallel}$), whose impact yields the total device impedance (Z_{Device}) represented by the purple dot (5). The device inductance is not taken into account in this section of the analysis since it becomes part of the transformer model. Note that if the DC load resistance is chosen to have a very high value, then the device impedance will result in the complex conjugate of the empty device inductance (L_{Coil_D}).

Transformer Reflected Impedance – Single Device

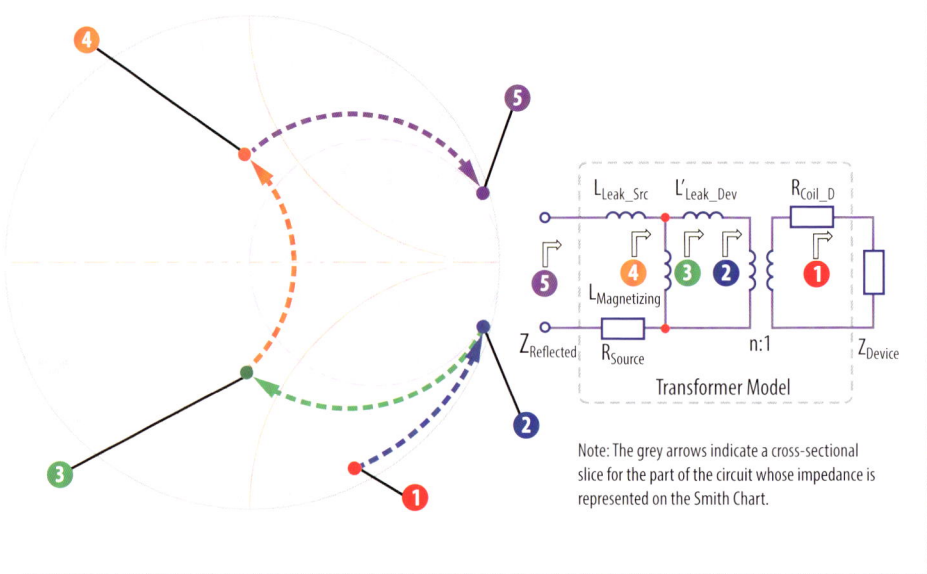

Note: The grey arrows indicate a cross-sectional slice for the part of the circuit whose impedance is represented on the Smith Chart.

Transformer Reflected Impedance – Single Device

Having determined the device impedance loading (Z_{Device}) on the transformer, it can now be transformed to the source side. Starting with the device impedance, the coil device side losses (R_{Coil_D}) are added. This results in a very small shift in impedance in most cases, and is required for high efficiency in wireless power applications.

To transform the device impedance to the source side, the impedance is multiplied by the square of the turns-ratio ($Z_{Device} \cdot n^2$) to reveal the device impedance on the source side, shown by the blue dot (2). Next, the device side transformed to the source side leakage inductance (L'_{Leak_Dev}) is added resulting in the green dot (3). This is connected in parallel with the magnetizing inductance ($L_{Magnetizing}$) and yields the orange dot (4).

Lastly, the series components, the source side leakage inductance (L_{Leak_Src}), and source coil winding resistance (R_{Source}) are added to yield the total reflected impedance ($Z_{Reflected}$) as shown by the purple dot (5).

Tuned Reflected Impedance

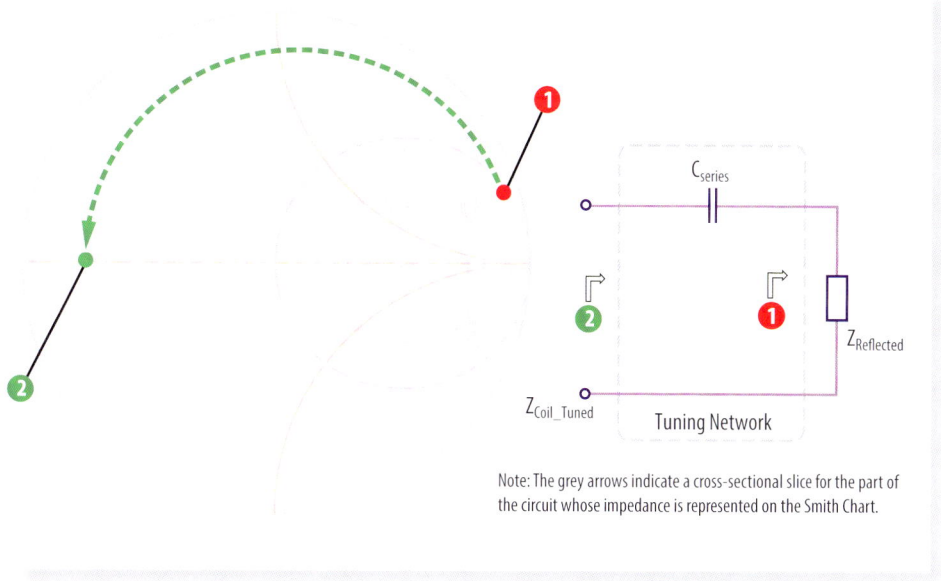

Note: The grey arrows indicate a cross-sectional slice for the part of the circuit whose impedance is represented on the Smith Chart.

Tuned Reflected Impedance

Having calculated the total reflected impedance ($Z_{Reflected}$), shown by the red dot (1), it can be used to determine the total tuned coil-set impedance (Z_{Coil_Tuned}) shown by the green dot (2). In this case, the series capacitance (C_{series}) is added to the total reflected impedance.

In a good design, using an applicable DC load resistance, the result should still be very close to the real axis of the Smith Chart, which indicates that the coil set remains on resonance. Caution should be exercised if no real impedance shift is observed for low DC resistance values, for this would indicate that the device may not be correctly tuned, or the coupling factor is too low.

Effect of Device Load on Reflected Impedance

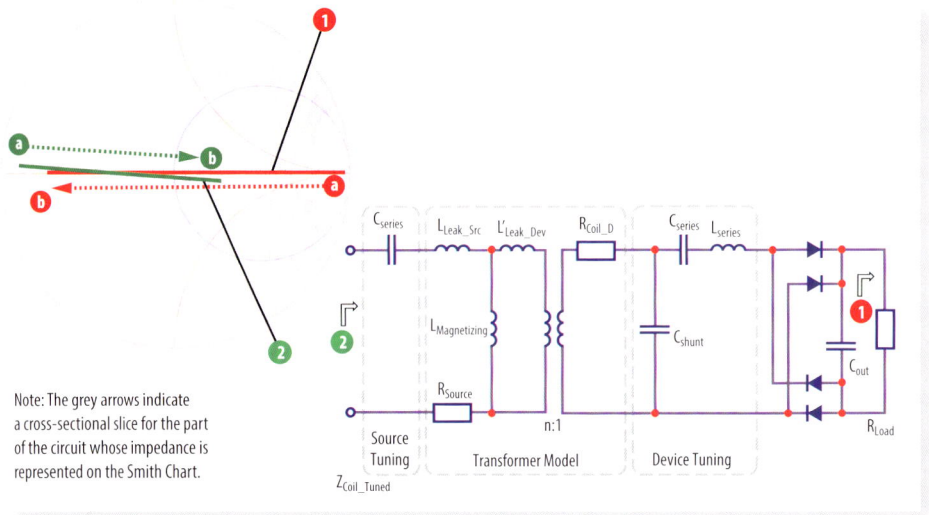

Note: The grey arrows indicate a cross-sectional slice for the part of the circuit whose impedance is represented on the Smith Chart.

Effect of Device Load on Reflected Impedance

The analysis method used to this point can now be used to determine the impact of the DC load resistance (R_{Load}) variation, which would occur by a charging battery, on the impedance of the tuned coil-set (Z_{Coil_Tuned}).

Starting with a high DC load resistance value shown by the red dot (a) and decreasing until it reaches (b), the reflected impedance of the tuned source coil is shown by the green trace and moves in the opposite direction starting with a low impedance at (a) and increasing to end at (b). This is the negative impedance effect of the highly resonant wireless power coil-set.

Effect of Coupling on Reflected Impedance

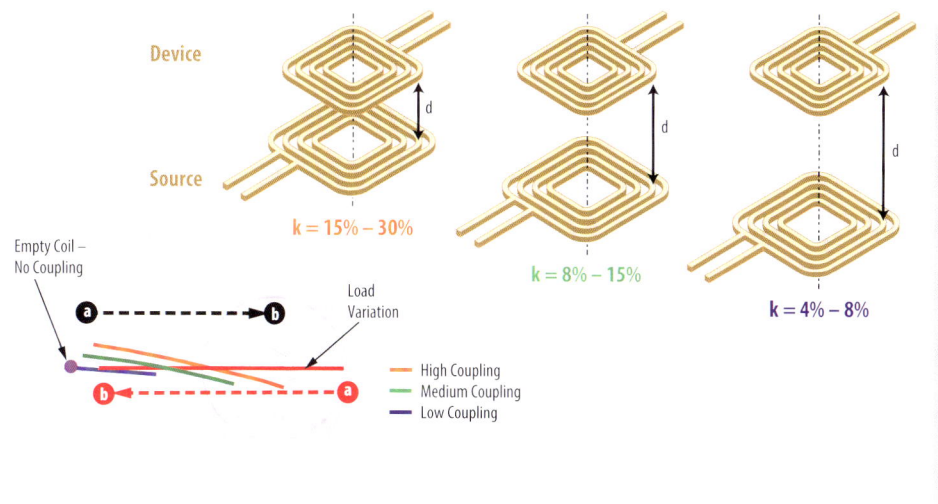

Effect of Coupling on Reflected Impedance

The analysis methodology can further be used to determine the impact of coupling factor (k) on the tuned source coil reflected impedance (Z_{Coil_Tuned}).

As in the previous example, the DC load resistance is varied starting at a high DC load resistance value shown by the red dot (a) and is decreased until it reaches (b). When the device coil is placed close to the source coil it yields a high coupling of between 15% and 30%, as shown on the left image. The high coupling results in a large tuned source coil reflected impedance variation, as shown by the orange trace. Again, the impedance shift is reversed going from the low impedance, which is indicated by the black dot (a) and increasing in the direction of the black dot (b).

Increasing the separation between the device and source coil, shown by the center image, will decrease the coupling between them. At between 8% and 15% coupling, the variation in the tuned source coil reflected impedance decreases, which is indicated by the green trace. Additional increase in separation distance between the device and source coil, shown by the right image, will further decrease the coupling down to between 4% and 8% with a corresponding decrease in the tuned source coil reflected impedance as shown by the blue trace.

Complete removal of the device coil yields the purple dot, which is the original tuned set-point for the source coil.

The traces have been offset for clarity, they actually lie one on top of the other.

Coupling Adjustment for Targeted Impedance Variation

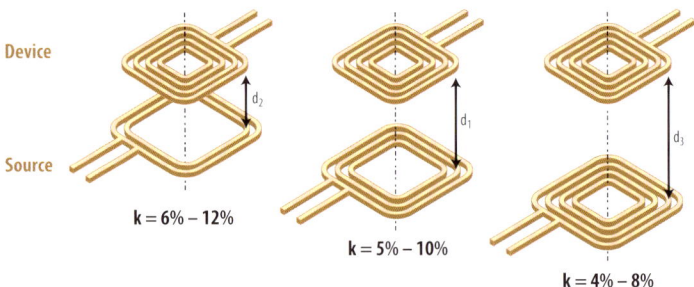

Coupling Adjustment for Targeted Impedance Variation

The analysis methodology can further be used as a design tool for matching the coil-set with a specific use case specification. Changes can be made to the coil-set to avoid a too large or too small tuned source coil reflected impedance variation.

In most applications, the device requirements are set based on the A4WP consortium's approval process and the requirements for compatibility between conforming systems. Therefore only changes to the source coil can be made to match it to a particular impedance specification. If designing both the device and source coils, then both designs can be adjusted. In this discussion, the device design will remain fixed to better illustrate this concept.

Starting with the center image case, where the coupling factor lies between 5% and 10% for a device to coil separation distance of d_1, the tuned source coil's reflected impedance would yield a specific impedance range. If that range is too high or too low, due to the requirement to meet either a separation distance of d_2 or d_3, then the number of turns on the source coil can be adjusted to correct the impedance range.

Removing turns from the source coil effectively reduces the coupling coefficient if the separation distance is maintained, as shown in the left image. Adding turns to the source coil effectively increases the coupling coefficient if the separation distance is maintained, as shown in the right image. By adjusting the distance to the use case specification, given by d_2 or d_3 respectively, the resulting coupling coefficient will remain largely the same. This will become an important design tool when trying to match the coil impedance to work efficiently with a particular amplifier.

Amplifier Drive Capability

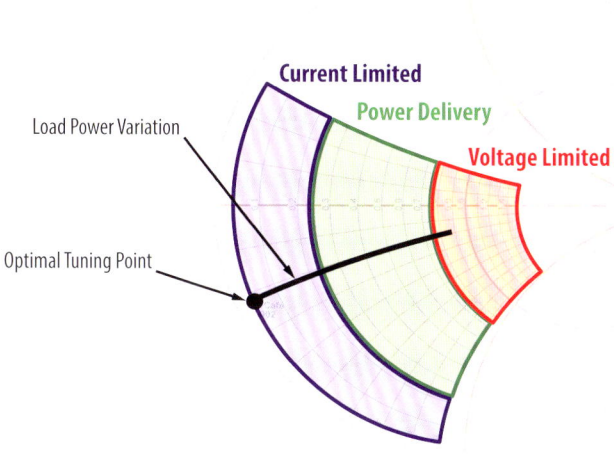

Amplifier Drive Capability

Now that the coil-set impedance can easily be analyzed, the impact of the impedance variation on a potential amplifier needs to be understood. From the analysis, it was determined that the DC load resistance variation results in an inverse impedance variation, where the lowest DC load resistance results in the highest reflected impedance.

If the coil design yields a low value for the maximum reflected impedance, the amplifier needs to operate in the blue region. Operation in this region requires a high current and it will be more difficult to deliver high power.

If the coil design yields a high value for the maximum reflected impedance, the amplifier needs to operate in the red region. Operation in this region requires a high voltage and again will prove difficult to deliver high power.

A balance needs to be found for the coil design to find the optimal operating point that matches the amplifier capability, as shown by the green region. This will yield the highest efficiency for both the coil-set and the amplifier based on the power delivery requirements.

Furthermore, amplifiers will have a reactive impedance drive capability, which needs to be determined and matched to the coil-set. If the coil-set reactive impedance variation falls within the amplifier capability, then the coil set tuning should target the center of the reactive impedance range, as shown by the black dot. This only requires adjustment of the source coil series tuning capacitance. Expected impedance variations will then completely fall inside the amplifier capability, and DC load resistance variation will follow the black trace when the coil is operated under the conditions at which it was tuned.

CHAPTER 3:

Wireless Power Amplifier Topologies Overview

Topologies Suitable for Wireless Power Transfer Amplifiers

- Class E
- Current-mode class D
- Voltage-mode class D
- ZVS Voltage-mode class D
- Differential-mode variations

Wireless Power Transfer Amplifier Topologies Overview

In this section we present various amplifier topologies suitable for wireless power transfer. The main selection criteria for any topology are high operating efficiency and reasonable cost. As a result, most traditional RF amplifier topologies are not considered for use in wireless power systems since their efficiencies are too low. Only switching-based amplifier topologies that make use of soft-switching techniques are suitable candidates.

The most popular topology for this application is the class E amplifier, as it has a theoretical 100% conversion efficiency [3.1]. Class D topologies are also good candidates, as these amplifiers exhibit very high conversion efficiency [3.2, 3.3]. Both the voltage- and current-mode class D amplifier topologies are covered in this handbook.

A variation of the traditional voltage-mode class D amplifier that is well suited for wireless power is the zero voltage switching (ZVS) voltage-mode topology [3.4] which is also presented.

Finally, some of these amplifier topologies can operate in a differential mode and are also discussed.

[3.1] F. H. Raab, "Idealized operation of the class E tuned power amplifier," *IEEE Transactions on Circuits and Systems*, Vol.24, No. 12, pp. 725–735, December 1977.

[3.2] D. Oliveira, C. Duartey, V. G. Tavaresy, P. G. de Oliveira, "Design of a Current-Mode Class-D Power Amplifier in RF-CMOS," *XXIV Conference proceeding on Design of Circuits and Integrated Systems (DCIS'09)*, November 2009, pp. 85–89.

[3.3] M. A. De Rooij and J. T. Strydom, "eGaN® FETs in Low Power Wireless Energy Converters," *Electro-Chemical Society Transactions on GaN Power Transistors and Converters*, Vol. 50, No. 3, pp. 377–388, October 2012.

[3.4] A. Lidow, M. A. de Rooij, "Performance Evaluation of Enhancement-Mode GaN transistors in Class-D and Class-E Wireless Power Transfer Systems," *Bodo's Power Systems*, May 2014, pp. 56–60.

Coil Network Simplification

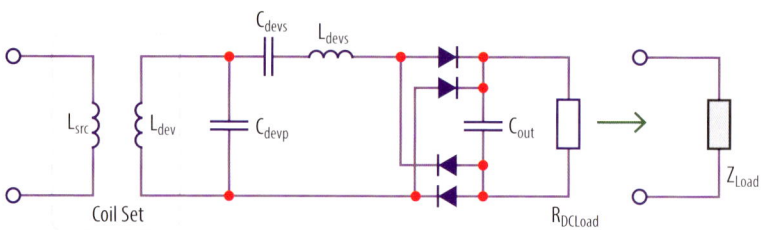

Coil Network Simplification

The entire coil system can be simplified into an equivalent impedance (Z_{Load}) that can be used to design, evaluate, and compare various topologies and devices. This simplification requires accurate impedance values for all the components in the equivalent circuit at the operating frequency, including the equivalent transformer circuit model discussed in the previous section.

The impedances can be determined from s-parameters provided by the manufacturer, or alternatively, by s-parameter measurement of the discrete components using a VNA. The rectifier poses a challenge, as neither s-parameter data nor VNA measurements can be used since the magnitude of the VNA signal in most cases is too small to effectively activate the diodes.

Whenever the Z_{Load} is discussed later in this work, the full circuit shown above is being described and is referred to as the reflected load impedance. Whenever R_{DCLoad} is discussed, then only the final DC load resistance is being described.

Class E Amplifier

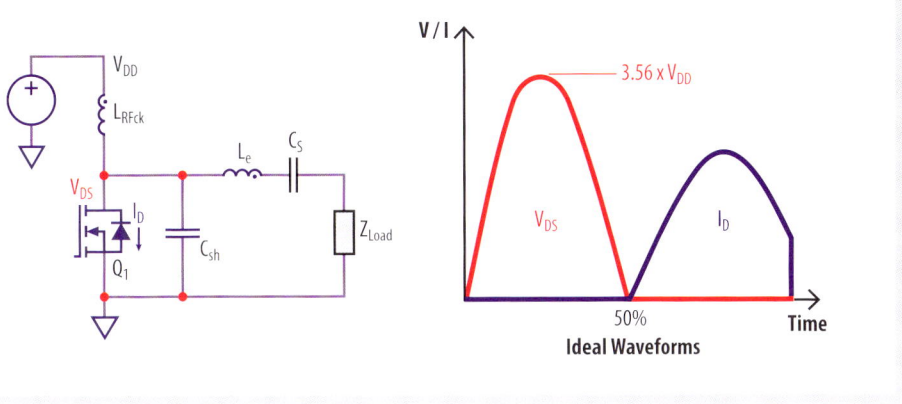

Ideal Waveforms

Class E Amplifier

Shown here is the single-ended class E amplifier together with the ideal operating waveforms [3.5]. Due to ISM band restrictions, the amplifier operates with a fixed frequency and 50% fixed duty cycle. Under ideal operating conditions the device voltage rating can be 3.56 times the supply voltage (V_{DD}). However, due to load and coupling variations, the peak voltage across the device can be as high as 7 times the supply voltage.

The amplifier designer must consider all operating conditions when selecting a suitable power transistor [3.5]. The output capacitance (C_{OSS}) of the transistor becomes part of the matching network design and is effectively absorbed into the matching network design. However, in some cases C_{OSS} can be too high, which can lead to very high device losses under certain load conditions. For this amplifier, the tuned coil RMS voltage, across the series combination of Z_{Load} and C_S is approximately 0.707 times the supply voltage.

One advantage of the class E topology is that it requires only a ground-referenced gate driver, which can reduce costs. Specifically, the LM5114 [3.6] or UCC27611 [3.7] from Texas Instruments can be used for larger EPC eGaN FETs. In the latest generation of eGaN FETs, such as the EPC2037 [3.8], the input capacitance is low enough that the controller's digital output is sufficient to operate the transistor, thus saving the cost, space, and power losses associated with the driver.

[3.5] M. A. de Rooij, "eGaN® FET based Wireless Energy Transfer Topology Performance Comparisons," *International Exhibition and Conference for Power Electronics, Intelligent Motion, Renewable Energy and Energy Management (PCIM Europe)*, May 2014, pp. 610–617.

[3.6] Texas Instruments, "Single 7.6A Peak Current Low-Side Gate Driver," LM5114 datasheet, Jan 2011 [Revised March 2013]. [Online] Available: http://www.ti.com/lit/ds/symlink/lm5114.pdf

[3.7] Texas Instruments, "4A/6A High-Speed 5-V Drive Optimized Single Gate Driver," UCC27611 datasheet, Dec 2012. [Online] Available: http://www.ti.com/lit/ds/symlink/ucc27611.pdf

[3.8] Efficient Power Conversion Corporation, "Enhancement Mode Power Transistor," EPC2037 datasheet, June 2015. [Online] Available: http://epc-co.com/epc/documents/datasheets/EPC2037_preliminary.pdf

Current-Mode Class D Amplifier

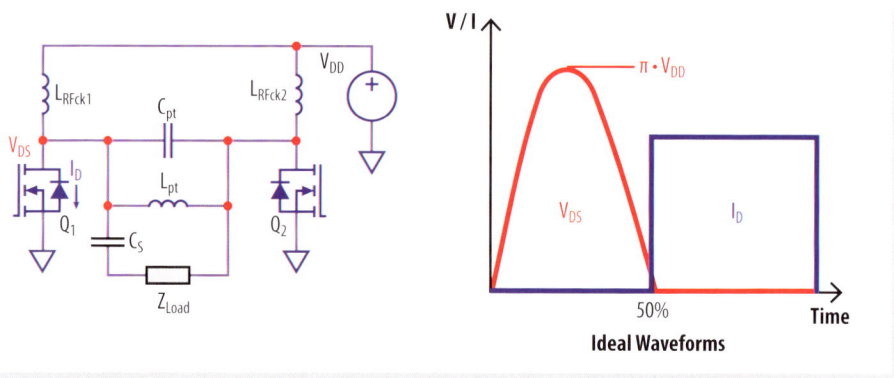

Ideal Waveforms

Current-Mode Class D Amplifier

Next is shown the ideal current-mode class D amplifier [3.9]. Due to ISM band restrictions, this amplifier operates with a fixed frequency and a 50% fixed duty cycle. Under ideal operating conditions, the power transistor's (FET) voltage rating is pi (π) times the supply voltage (V_{DD}) and, unlike the class E amplifier, it is largely independent of load conditions. The FET output capacitance (C_{OSS}) appears in parallel with the resonant tank circuit capacitance (C_{pt}), and thus becomes part of the matching network design, since it is effectively absorbed into the matching network.

One drawback of the current-mode class D amplifier is the high current in the resonant inductor that can lead to high losses. This can be mitigated by selecting an inductor with a high Q factor at the operating frequency. For this amplifier, the tuned-coil RMS voltage, across the series combination of Z_{Load} and C_S, is approximately 2.22 times the supply voltage, giving it a very high voltage gain and thus making it a suitable candidate when the supply voltage is limited.

Another advantage of the current-mode class D amplifier is that it requires only a ground-referenced gate driver. In particular, the LM5114 [3.10] or, UCC27611 [3.11] from Texas Instruments can be used for larger EPC eGaN FETs. In the latest generation of eGaN FETs, such as the EPC2037 [3.12], the input capacitance is low enough that the controller's digital output is sufficient to operate the transistor, thus saving the cost, space, and power losses associated with the driver.

[3.9] M. A. de Rooij, "eGaN® FET based Wireless Energy Transfer Topology Performance Comparisons," *International Exhibition and Conference for Power Electronics, Intelligent Motion, Renewable Energy and Energy Management (PCIM - Europe)*, May 2014, pp. 610–617.

[3.10] Texas Instruments, "Single 7.6A Peak Current Low-Side Gate Driver," LM5114 datasheet, Jan 2011 [Revised March 2013]. [Online] Available: http://www.ti.com/lit/ds/symlink/lm5114.pdf

[3.11] Texas Instruments, "4A/6A High-Speed 5-V Drive Optimized Single Gate Driver," UCC27611 datasheet, Dec 2012. [Online] Available: http://www.ti.com/lit/ds/symlink/ucc27611.pdf

[3.12] Efficient Power Conversion Corporation, "Enhancement Mode Power Transistor," EPC2037 datasheet, June 2015. [Online] Available: http://epc-co.com/epc/documents/datasheets/EPC2037_preliminary.pdf

Voltage-Mode Class D Amplifier

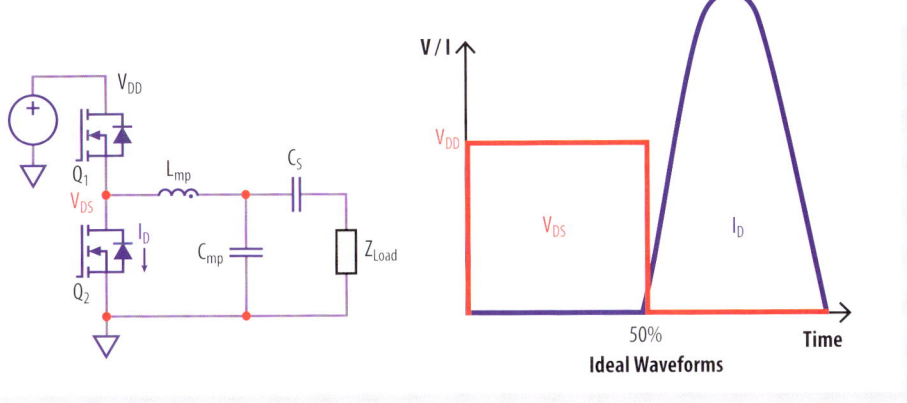

Ideal Waveforms

Voltage-Mode Class D Amplifier

The traditional voltage-mode class D amplifier is presented here [3.13]. Due to ISM band restrictions, this amplifier operates with a fixed frequency and a 50% fixed duty cycle. Under the ideal operating conditions shown, it uses zero current switching (ZCS) to establish high operating efficiency. However, operation at high frequency still requires the switching devices to transition voltage. It should be noted that devices with a high output capacitance (C_{OSS}) can have significant losses and reduce the system efficiency. For wireless power transfer operating at 6.78 MHz it is therefore necessary to change the operation of this class D amplifier to zero voltage switching (ZVS), which is achieved by tuning the load to appear inductive to the amplifier.

The current in the resultant inductance is used to self-commutate the switch-node voltage transitions, yielding a significant improvement in amplifier efficiency. However, tuning the load to appear inductive to the amplifier at the operating frequency shifts the resonant frequency below the tuned operating frequency, and hence decreases the coil system efficiency.

Due to the amplifier's configuration, the voltage rating of the devices can be the same as the supply voltage (V_{DD}). However, in some extreme and rare load conditions, it may be necessary to add some margin, but no more than 25% if using the optimal layout technique [3.14], for voltage ringing. Although, for this amplifier the tuned coil RMS voltage across the series combination of Z_{Load} and C_S is dependent on the magnitude of the supply voltage V_{DD}. The ratio of V_{DD} with respect to the tuned coil RMS voltage is dependent on the gain of the matching network (L_{mp} and C_{mp}).

The voltage-mode class D amplifier requires a level-shifting gate driver and the LM5113 [3.15] can therefore be used to drive the eGaN FETs.

[3.13] M. A. de Rooij, "eGaN® FET based Wireless Energy Transfer Topology Performance Comparisons," *International Exhibition and Conference for Power Electronics, Intelligent Motion, Renewable Energy and Energy Management (PCIM Europe)*, May 2014, pp. 610–617.

[3.14] D. Reusch, J. Strydom, "Understanding the Effect of PCB Layout on Circuit Performance in a High Frequency Gallium Nitride Based Point of Load Converter," *Applied Power Electronics Conference, APEC 2013*, pp. 649–655, 16–21 March 2013.

[3.15] Texas Instruments, "LM5113 5A, 100V Half-Bridge Gate Driver for Enhancement Mode GaN FETs," LM5113 datasheet, Apr. 2013. [Online] Available: www.ti.com/product/lm5113

ZVS Voltage-Mode Class D Amplifier

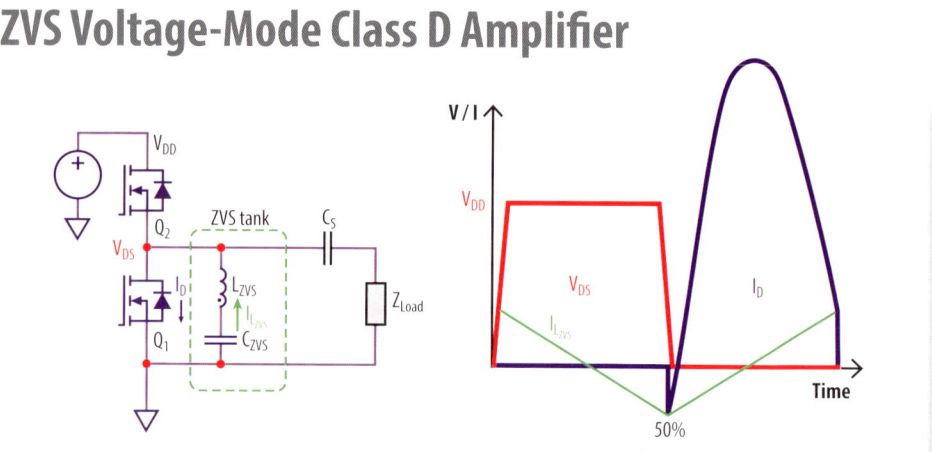

Ideal Waveforms

ZVS Voltage-Mode Class D Amplifier

The single-ended ZVS voltage-mode class D amplifier together with the ideal operating waveforms is shown here [3.16]. Due to ISM band restrictions, this amplifier operates with a fixed frequency and 50% fixed duty cycle. The circuit includes a non-resonant ZVS tank circuit that is used to self-commutate the switch-node voltage transitions and effectively the C_{OSS} of the devices. This allows the converter to operate as a no-load buck converter with ZVS transitions.

Due to its configuration, the voltage rating of the devices is always the same as the supply voltage (V_{DD}). However, in some extreme and rare load conditions, it may be necessary to add some margin, although no more than 25%, if using the optimal layout technique [3.17] for voltage ringing.

Another advantage of this topology is that the ZVS tank circuit does not carry any load current and, as such, will have very low operating losses. For this amplifier, the tuned coil RMS voltage across the series combination of Z_{Load} and C_S is approximately 0.45 times the supply voltage.

The ZVS voltage-mode class D amplifier requires a level-shifting gate driver for which the LM5113 [3.18] can be used to drive eGaN FETs.

[3.16] M. A. de Rooij, "eGaN® FET based Wireless Energy Transfer Topology Performance Comparisons," *International Exhibition and Conference for Power Electronics, Intelligent Motion, Renewable Energy and Energy Management (PCIM Europe)*, May 2014, pp. 610–617

[3.17] D. Reusch, J. Strydom, "Understanding the Effect of PCB Layout on Circuit Performance in a High Frequency Gallium Nitride Based Point of Load Converter," *Applied Power Electronics Conference, APEC 2013*, pp. 649–655, 16–21 March 2013.

[3.18] Texas Instruments, "LM5113 5A, 100V Half-Bridge Gate Driver for Enhancement Mode GaN FETs," LM5113 datasheet, Apr. 2013. [Online] Available: www.ti.com/product/lm5113

Differential-Mode Class E Amplifier

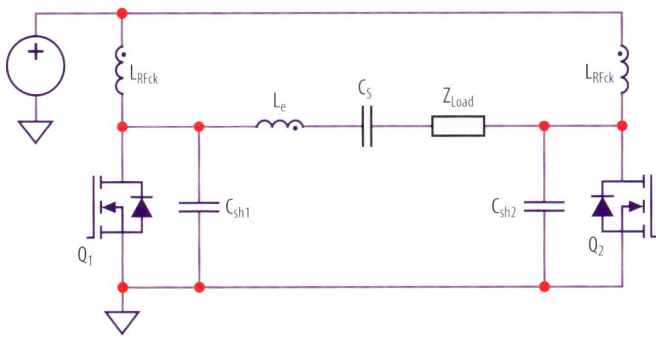

Differential-Mode Class E Amplifier

The class E amplifier can be configured in differential-mode. In this case, two single-ended versions of the amplifier are connected back-to-back, with the load appearing between the outputs of each single-ended amplifier and being operated differentially.

The extra inductor (L_e) of each amplifier can be combined into a single element for simplification. The differential-mode amplifier essentially doubles the output power capability over the single-ended version when it is operating with the same load and supply voltage (V_{DD}). As in the case of the single-ended version, the peak device voltage rating is 3.56 times the supply voltage (V_{DD}), but can again be as high as 7 times, subject to the load and coupling variations of the load.

Device selection for the differential-mode class E amplifier is the same as for the single-ended case, where high C_{OSS} can lead to very high device losses under certain load conditions. For this amplifier the tuned coil RMS voltage across the series combination of Z_{Load} and C_S is now doubled, at approximately 1.414 times the supply voltage. Another advantage of this amplifier is that it requires only a ground-referenced gate driver. The LM5113 [3.19] or two UCC27611s [3.20] can be used in conjunction with larger EPC eGaN FETs. In the latest generation of eGaN FETs, such as the EPC2037 [3.21], the input capacitance is low enough that the controller's digital output is sufficient to operate the transistor, thus saving the cost, space, and power losses associated with the driver.

[3.19] Texas Instruments, "LM5113 5A, 100V Half-Bridge Gate Driver for Enhancement Mode GaN FETs," LM5113 datasheet, Apr. 2013. [Online] Available: www.ti.com/product/lm5113

[3.20] Texas Instruments, "4A/6A High-Speed 5-V Drive Optimized Single Gate Driver," UCC27611 datasheet, Dec 2012. [Online] Available: http://www.ti.com/lit/ds/symlink/ucc27611.pdf

[3.21] Efficient Power Conversion Corporation, "Enhancement Mode Power Transistor," EPC2037 datasheet, June 2015. [Online] Available: http://epc-co.com/epc/documents/datasheets/EPC2037_preliminary.pdf

Differential-Mode ZVS Class D Amplifier

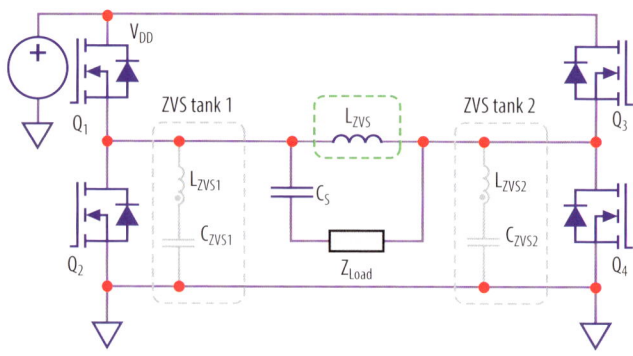

Differential-Mode ZVS Class D Amplifier

The ZVS class D can be configured as a differential-mode version. In this case, two single-ended versions of the amplifier are connected back-to-back, with the load appearing between the outputs of each single-ended amplifier and operated differentially.

The differential-mode ZVS class D amplifier can have two possible configurations. The configuration choice depends on the operating specifications. In the first configuration, two individual single-ended ZVS class D amplifiers are used to drive the load differentially. This configuration yields higher tolerance to component variations that can improve timing-related efficiency but increases component count. It is also more suited in applications requiring additional control such as On/Off Key (OOK) modulation. In the second configuration, the ZVS inductor (L_{ZVS}) is shared between the two halves and are added together for a single inductor and, as a consequence, the ZVS capacitors (C_{ZVSx}) can be eliminated. The differential-mode amplifier essentially doubles the output power capability over the single-ended version when operating with the same load and supply voltage (V_{DD}).

As in the case of the single-ended version, the device voltage rating is again the supply voltage (V_{DD}). However, in some extreme and rare load conditions, it may be necessary to add some design margin, but no more than 25% if using the optimal layout [3.22] for voltage ringing. Again, the ZVS tank circuit does not carry any load current and will have very low operating losses.

For this amplifier, the tuned coil RMS voltage across the series combination of Z_{Load} and C_S is approximately 0.9 times the magnitude of the supply voltage. This amplifier requires a level-shifting gate driver and therefore two LM5113s [3.23] can be used for driving eGaN FETs.

[3.22] D. Reusch, J. Strydom, "Understanding the Effect of PCB Layout on Circuit Performance in a High Frequency Gallium Nitride Based Point of Load Converter," *Applied Power Electronics Conference, APEC 2013*, pp. 649–655, 16–21 March 2013.
[3.23] Texas Instruments, "LM5113 5A, 100V Half-Bridge Gate Driver for Enhancement Mode GaN FETs," LM5113 datasheet, Apr. 2013. [Online] Available: www.ti.com/product/lm5113

What is Needed to Design the Optimal Wireless Power Transfer Amplifier?

Coil impedance range:

- Determines coil voltage and current for desired load power
- Affected by DC load power range
- Depends on coupling factor between the source and device coils, or
- Design based on a standard, e.g., A4WP

What is Needed to Design the Optimal Wireless Power Transfer Amplifier?

Before an amplifier can be designed for wireless power transfer, some basic information on the overall system is needed, including the real component of the reflected (tuned source coil with device present) impedance range that the amplifier will need to drive. This information can be used to determine the coil voltage and current based on the specific load power required, which ultimately drives the design of the amplifier.

The real component of the entire reflected impedance range should include the DC load variation and coupling changes between the device and source coils. Alternatively, a standard, such as the A4WP [3.24] that provides the real component of the reflected impedance range, can be used.

[3.24] *A4WP PTU Resonator Class 3 Design - Spiral Type 210–140 Series (PTU 3-0001)*, A4WP standard document RES-14-0008 Ver. 1.2, June 30, 2014.

Design of a Class E Amplifier

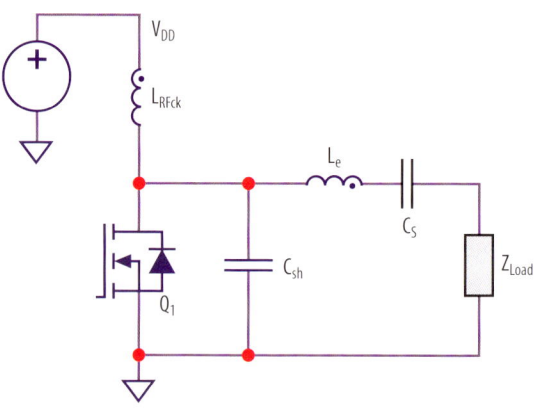

Design of a Class E Amplifier

The first amplifier design that we will cover is the single-ended class E. This is the most complex of the amplifier topologies due to its double resonant frequency structure. For this amplifier only three components need to be specifically designed: 1) the extra inductor (L_e), 2) the shunt capacitor (C_{sh}) and, 3) the selection of a suitable switching device. The RF choke (L_{RFck}) value is less critical and hence can be chosen or designed.

The design equations for the class E amplifier have been derived by N. Sokal [3.25]. To simplify these equations, the value of Q_L [3.25] is set to infinity, which is a reasonable approximation in this application. The design needs to have a specific reflected resistance (R_{Load}) value that is used to begin the design, which then drives the values of the other components, including the magnitude of the supply voltage.

[3.25] N.O. Sokal, "Class-E RF Power Amplifiers," *QEX*, Issue 204, pp. 9–20, January/ February 2001.

Impact of Load Resistance on the Class E FET

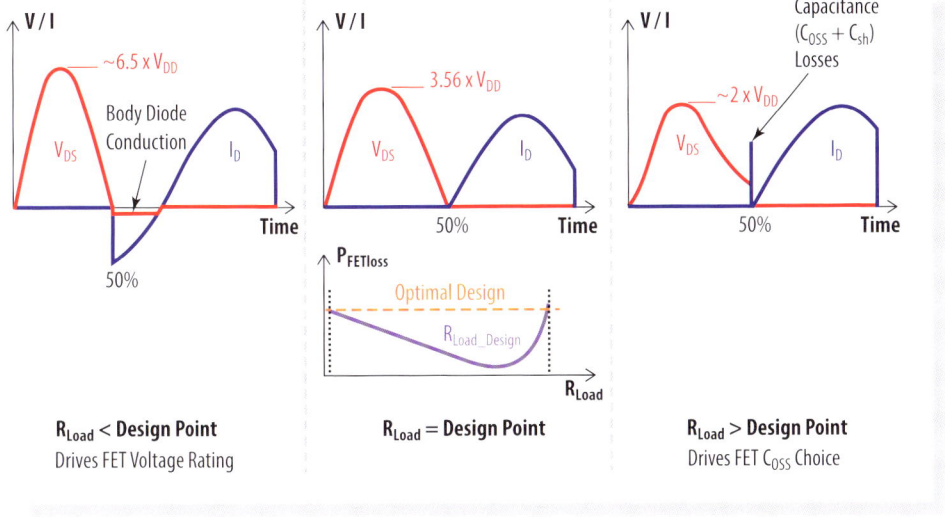

R_{Load} < Design Point
Drives FET Voltage Rating

R_{Load} = Design Point

R_{Load} > Design Point
Drives FET C_{OSS} Choice

Impact of Load Resistance on the Class E FET

The impact of reflected load resistance variation is significant to the performance of the class E amplifier, and must be carefully analyzed to select the optimal design resistance.

When operating a class E amplifier with a reflected load resistance (R_{Load}) that is below the design value (see the waveform on the left), the load tends to draw current from the amplifier too quickly. To compensate for this condition, the amplifier supply voltage is increased to yield the required output power. The shorter duration of the energy charge cycle leads to a significant increase in the voltage across the switching device. This is needed to capture sufficient energy and results in device body diode conduction during the remainder of the device off period. This period is characterized by a linear increase in device losses as a function of decreasing reflected load resistance (R_{Load}).

When operating the class E amplifier with a reflected load resistance (R_{Load}) that is above the design value (see the waveform on the right), the load tends to draw insufficient current from the amplifier, resulting in an incomplete voltage transition. When the device switches there is a residual voltage across the device, which leads to C_{OSS} losses. This period in the cycle is characterized by an exponential increase in device losses as a function of increasing reflected load resistance.

Given these two extremes of the operating reflected load resistance (R_{Load}), the optimal point between them must be determined. In this case, the optimal point yields the same device losses for each of the extreme reflected load resistance points and is shown in the lower center graph. This optimal design point can be found through trial and error, or using circuit simulation.

Class E Amplifier Passive Component Design

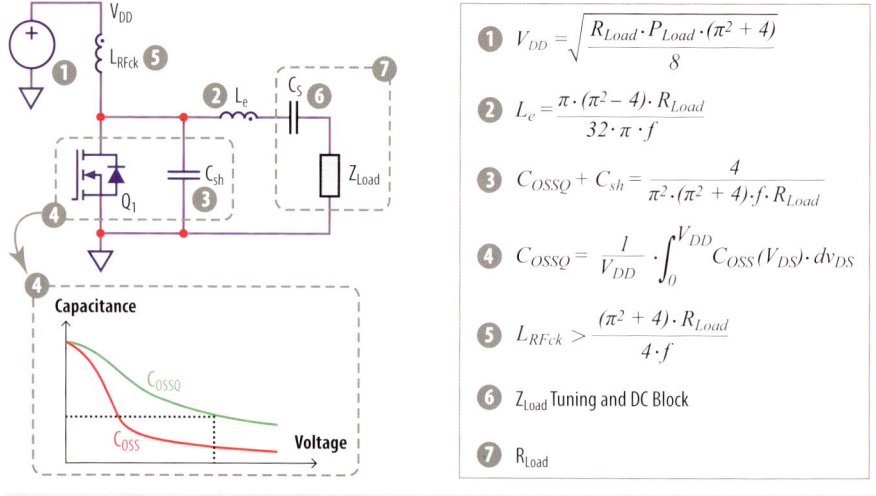

$$V_{DD} = \sqrt{\frac{R_{Load} \cdot P_{Load} \cdot (\pi^2 + 4)}{8}}$$

$$L_e = \frac{\pi \cdot (\pi^2 - 4) \cdot R_{Load}}{32 \cdot \pi \cdot f}$$

$$C_{OSSQ} + C_{sh} = \frac{4}{\pi^2 \cdot (\pi^2 + 4) \cdot f \cdot R_{Load}}$$

$$C_{OSSQ} = \frac{1}{V_{DD}} \cdot \int_{0}^{V_{DD}} C_{OSS}(V_{DS}) \cdot dv_{DS}$$

$$L_{RFck} > \frac{(\pi^2 + 4) \cdot R_{Load}}{4 \cdot f}$$

⑥ Z_{Load} Tuning and DC Block

⑦ R_{Load}

Class E Amplifier Passive Component Design

The class E amplifier passive component design starts with the reflected load impedance resistance value (R_{Load}). The reactive component of the reflected impedance Z_{Load} is tuned out using C_S, which also serves to block DC. It is a common mistake to ignore the need for the DC block, where a failure to do so can yield a DC current from the supply through to the coil, and lead to additional losses in several components in that path.

First, using the equations in the figure, both the extra inductor (equation 2) and shunt capacitor (equation 3) values can be determined [3.26, 3.27]. The value of the shunt capacitor includes the C_{OSS} of the switching device, which must be subtracted from the calculated value to yield the actual external capacitor (C_{sh}) value. To do this, first the magnitude of the supply voltage (V_{DD}) is calculated using equation 1, which in turn can be used to determine the peak device voltage (3.56 * V_{DD}).

The RMS value of the peak device voltage is then used to determine the C_{OSSQ} of the device at that voltage. This is the capacitance that will be deducted from the calculated shunt capacitor to reveal the external shunt capacitor (C_{sh}) value. The C_{OSSQ} of the device can be calculated by integrating the C_{OSS} as a function of voltage using equation 4. If the C_{OSSQ} value is larger than the calculated shunt capacitance, then the design cannot be realized for the reflected load resistance specified and either a new device must be selected with lower C_{OSSQ}, or the reflected load resistance (R_{Load}) must be adjusted.

Finally, the choke (L_{RFck}) can be designed using equation 5 and, in this case, a minimum value is specified. Larger values yield lower ripple current, which can lead to a more stable operating amplifier. A too-high value will lead to increased operating losses.

[3.26] M. Kazimierczuk, "Collector amplitude modulation of the class E tuned power amplifier," *IEEE Transactions on Circuits and Systems*, June 1984, Vol.31, No. 6, pp. 543–549.

[3.27] Z. Xu, H. Lv, Y. Zhang, Y. Zhang, "Analysis and Design of Class E Power Amplifier employing SiC MESFETs," *IEEE International Conference on Electron Devices and Solid-State Circuits (EDSSC) 2009*, 25–27 December 2009, pp 28–31.

Design of a Current-Mode Class D Amplifier

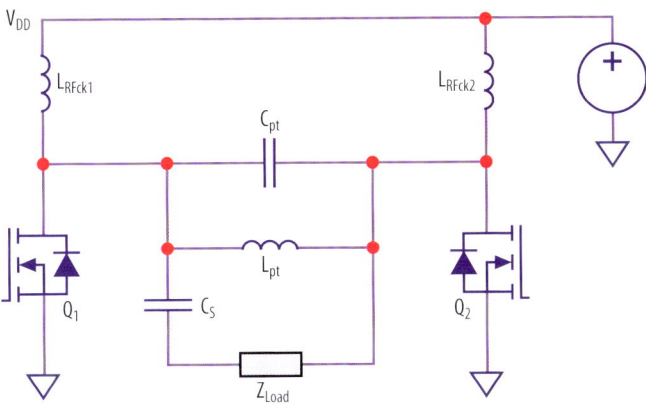

Design of a Current-Mode Class D Amplifier

The current-mode class D amplifier is the next amplifier that will be discussed. It is the simplest of the amplifiers to design, as it only requires the design of a resonant tank circuit. This amplifier requires only three components to be specified: 1) the resonant inductor (L_{pt}), 2) the resonant capacitor (C_{pt}) and 3) suitable switching devices. The RF chokes (L_{RFckX}) value are less critical and can be chosen or designed in a similar manner as for the class E amplifier.

This amplifier design does not need to have a specific reflected resistance (R_{Load}) value to begin the design since the resonant tank circuit (L_{pt} and C_{pt}) appears in parallel with the resonant load circuit (C_S and Z_{Load}). However, the quality factor (Q) of the resonant tank circuit should be high enough to overcome the reflected load resistance (R_{Load}) effects. The highest reflected load resistance in this case will drive the supply voltage (V_{DD}).

Current-Mode Class D Amplifier Passive Component Design

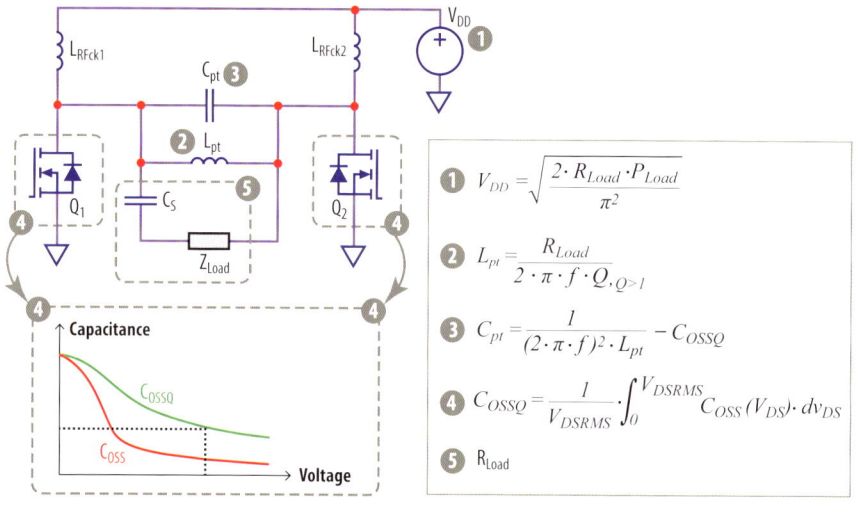

$$\text{1} \quad V_{DD} = \sqrt{\frac{2 \cdot R_{Load} \cdot P_{Load}}{\pi^2}}$$

$$\text{2} \quad L_{pt} = \frac{R_{Load}}{2 \cdot \pi \cdot f \cdot Q}, _{Q>1}$$

$$\text{3} \quad C_{pt} = \frac{1}{(2 \cdot \pi \cdot f)^2 \cdot L_{pt}} - C_{OSSQ}$$

$$\text{4} \quad C_{OSSQ} = \frac{1}{V_{DSRMS}} \cdot \int_0^{V_{DSRMS}} C_{OSS}(V_{DS}) \cdot dv_{DS}$$

$$\text{5} \quad R_{Load}$$

Current-Mode Class D Amplifier Passive Component Design

As with other amplifiers, the current-mode class D amplifier passive component design starts with the reflected load impedance resistance (R_{Load}) value. However, in this case the reflected load impedance resistance (R_{Load}) value is directly related to the power output (P_{Load}).

First, the reactive component of Z_{Load} is tuned out using C_S, which serves as a DC block similar to the case with the class E amplifier. The reflected load resistance (R_{Load}) can be used to calculate the required supply voltage (V_{DD}), using equation 1, for the amplifier. The balance of the design focuses on the resonant tank circuit (L_{pt} and C_{pt}), that is used to generate the coil voltage and can be calculated using equations 2 and 3.

The basic design of this amplifier is well documented in the literature [3.28]. By selecting a specific quality factor (Q) value for the resonant tank circuit, the value of the resonant inductor and capacitor can be determined. It is important to select a Q factor greater than one. If a Q factor of less than one is chosen, the losses in the inductor will become excessive and load variation will cause large shifts in resonant tank frequency, which will lead to increased device losses.

The device C_{OSS} will appear in parallel with the resonant tank circuit capacitance (C_{pt}) and needs to be included in the calculation. The C_{OSSQ}, using equation 4, is calculated in the same manner as for the class E case, but it is important to note that only one device is connected in parallel at any given time, and hence the C_{OSSQ} of only one device is ever added to the resonant tank circuit capacitance calculation.

[3.28] D. K. Choi, "High efficiency switched-mode power amplifiers for wireless communications," Ph.D. dissertation, University of California, Santa Barbara, CA, March 2001.

Design of a ZVS Class D Amplifier

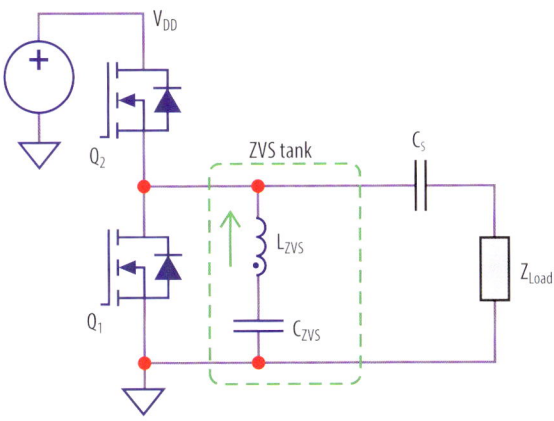

Design of a ZVS Class D Amplifier

The last amplifier design that will be covered is the ZVS class D. Here, only two components need to be designed, which are the ZVS inductor (L_{ZVS}) and the selection of suitable switching devices. The ZVS capacitor (C_{ZVS}) value is less critical and can be chosen or designed to a specific voltage ripple.

The ZVS class D amplifier is unaffected by reflected load resistance (R_{Load}) variation, as load current operates orthogonally to the ZVS tank circuit current. The highest reflected load resistance (R_{Load}) will determine the highest operating voltage for the circuit, as well as the device voltage rating. The required supply voltage and transition time of the switch-node then determines the design of the ZVS tank inductor.

ZVS Class D Amplifier Passive Component Design

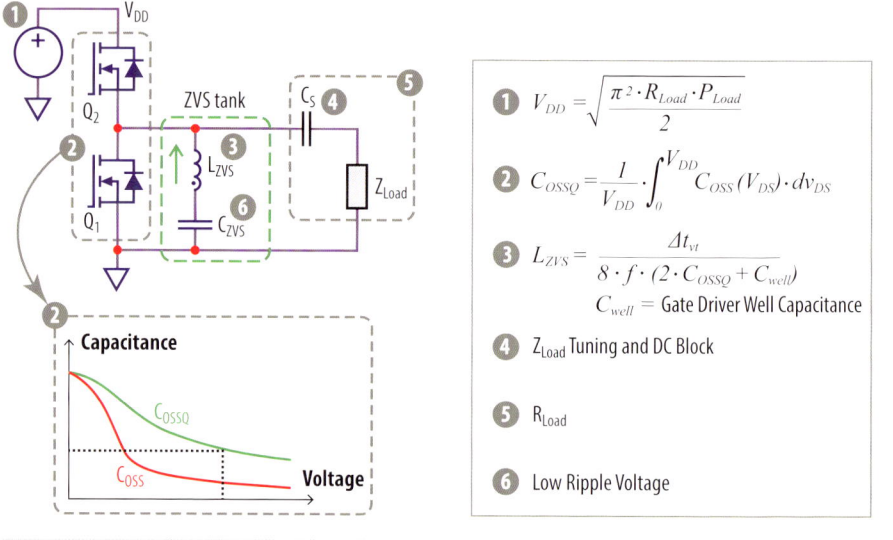

ZVS Class D Amplifier Passive Component Design

As in the case of the current-mode class D amplifier, the design of the ZVS class D amplifier begins with the reflected load resistance (R_{Load}) value and the required output power (P_{Load}). These parameters are used to calculate the required supply voltage (V_{DD}), using equation 1, which is the same as for the traditional voltage-mode class D design [3.29, 3.30].

Using the required supply voltage, suitable devices can be selected and their C_{OSSQ} determined using equation 2. The voltage used for C_{OSSQ} is the supply voltage (V_{DD}), and is calculated in the same manner as the previously discussed amplifiers.

Next, a switch-node transition time (Δt_{vt}) needs to be selected. A "too-long time" will degrade the performance of the amplifier with excessive effective duty cycle loss, and reduced immunity to imaginary reflected load impedance variations. A "too-short time" will increase operating losses for both the devices and the ZVS inductor. The combination of both devices' C_{OSSQ}, gate driver well capacitance (C_{well}) [3.31], operating frequency (f), and transition time is used to determine the ZVS tank inductance (L_{ZVS}) using equation 3. The gate driver well capacitance is included in the calculation, as it is typically of the same order of magnitude as the devices.

[3.29] D. K. Choi, "High efficiency switched-mode power amplifiers for wireless communications," Ph.D. dissertation, University of California, Santa Barbara, CA, March 2001.

[3.30] S-A. El-Hamamsy, "Design of high-efficiency RF class D power amplifier," *IEEE Transactions on Power Electronics*, Vol. 9, No. 3, May 1994, pp. 297–308

[3.31] N. Mehta, "Design Considerations for LM5113 Advanced GaN FET Driver During High-Frequency Operation," Texas Instruments Application Report SNVA723, November 2014, [Online] Available: www.ti.com/lit/an/snva723/snva723.pdf.

Wireless Power Amplifier Topology Overview Summary

Amplifier Parameter	Single-Ended Class E	Current-Mode Class D	Single-Ended Voltage-Mode Class D	Single-Ended ZVS Class D
# of FETs	1	2	2	2
Gate Driver Requirement	Ground Referenced	Ground Referenced	Level Shifted	Level Shifted
DC Supply to Coil RMS Voltage Gain	$= \frac{1}{\sqrt{2}} = 0.707$	$= \frac{\pi}{\sqrt{2}} = 2.22$	$= \frac{\sqrt{2}}{\pi} \cdot G_{Match}$	$= \frac{\sqrt{2}}{\pi} = 0.45$
Differential-Mode Configurable	Yes	Already	Yes	Yes
ZVS using Resonance	Yes	Yes	No (Load Dependent)	No
Load Current in Passives	Yes	No	Yes	No

Wireless Power Amplifier Topology Overview Summary

This table summarizes the differences between the various amplifier topologies presented.

Looking at the number of FETs used in each of the topologies, the higher the FET count the higher the costs can add up even though eGaN FET costs for wireless power have come down dramatically and rival the cost of power MOSFETs of equivalent voltage and on-resistance rating. Also, the level shifting requirements for the gate driver can add cost if integrated solutions are not available. Based on these criteria, the class E can be considered a good choice.

The DC supply voltage to tuned-coil RMS voltage gain is also given. A too-high gain will lead to high supply current with associated losses, and a too-low voltage gain will require high voltage to operate the amplifier. A correct balance will lead to lowest overall amplifier losses. This remains, however, an incomplete picture due to the impact of load variation that can shift losses. We will cover this in more detail later.

Some of the topologies can be configured as differential mode versions. This can be used as an effective method to double the power capability in a simple step.

Finally, resonance is used to establish zero voltage switching for some of these converters making them more susceptible to load variations, in particular where load current flows in those passive elements. In this case, the ZVS class D amplifier does not use resonance to establish ZVS or have load current in the ZVS circuit, making it much less susceptible to load variations as will be experimentally verified.

CHAPTER 4:

FET Selection for Highest Efficiency

Use a Figure of Merit (FOM) for selecting devices:
- Easy to compare technologies – GaN versus Silicon
- Straightforward to use

High frequency capability requires:
- Low gate capacitance
- Lowest practical drain capacitance
- No, or very low reverse recovery charge

FET Selection for Highest Efficiency

Having presented various wireless power transfer capable amplifier topologies, suitable devices for use with these amplifiers need to be selected. As discussed earlier, various technologies exist today and making the optimal choice for a transistor is very important.

This choice should give the highest possible amplifier efficiency and maximize wireless power transfer performance. The simplest and quickest method for selecting the best power devices for use in a wireless power transfer application is a Figure of Merit (FOM) approach [4.1]. The FOM used should encompass as many of the required characteristics for a suitable switching device in this application as possible. These device characteristics include low gate capacitance, low drain capacitance and low, or no, reverse recovery charge.

[4.1] D. Reusch, J. Strydom, "Evaluation of Gallium Nitride Transistors in High Frequency Resonant and Soft-Switching DC-DC Converters," *IEEE Applied Power Electronics Conference (APEC)*, 16–20 March 2014, pp. 464–470.

Figure of Merit for Evaluating FETs in Wireless Power Transfer Applications

With a ZVS transition, device gate-drain Q_{GD} charge is not important

MOSFET Q_{RR} is omitted as it leads to such high losses that high frequency operation is not possible

Although C_{OSS} is "absorbed" in the matching network it is still important because it:

- Drives off-resonance losses
- Determines design ability

$$FOM_{WPT} = R_{DS(on)} \bullet (Q_G - Q_{GD} + Q_{oss})$$

Figure of Merit for Evaluating FETs in Wireless Power Transfer Applications

Figures of merit are typically used to quickly and simply compare various power devices based upon different technologies in various topologies [4.2]. This simple FOM tool can be adopted for wireless power transfer applications, as well.

All the topologies to be evaluated will be ZVS, due to the high frequency nature of the circuits, and ZVS being the only reliable low-loss mechanism available. These circuits will need to operate close to ZVS conditions to maintain device losses within reasonable limits, even when operating off resonance and under non-ideal conditions. Since the device only switches at or near zero volts, the charge associated with gate-drain switching (Q_{GD}) is not significant and can be subtracted from the device's input charge (Q_G).

The magnitude of C_{OSS} is an important factor despite being absorbed into the matching circuits to establish ZVS, as it drives off-resonance losses and design-ability of the amplifier, most notably for the class E amplifier.

The reverse recovery of the silicon MOSFETs (Q_{RR}) is omitted as this condition leads to such high losses that high frequency operation is no longer possible. This is due to the reverse conduction prior to a hard switching event that includes full supply voltage C_{OSS} losses.

Based on these assumptions the Wireless Power Transfer (WPT) figure of merit for the evaluation of transistors can be defined simply as the on-state resistance ($R_{DS(on)}$) of the device multiplied by the total gate charge (Q_G) minus the voltage transition component (Q_{GD}) plus the output capacitance charge Q_{OSS} [4.3]. In other words:

$$FOM_{WPT} = R_{DS(on)} \bullet (Q_G - Q_{GD} + Q_{OSS})$$

[4.2] D. Reusch, J. Strydom, "Evaluation of Gallium Nitride Transistors in High Frequency Resonant and Soft-Switching DC-DC Converters," *IEEE Applied Power Electronics Conference (APEC)*, 16–20 March 2014, pp. 464–470.

[4.3] M. A. de Rooij, "Performance Comparison for A4WP Class-3 Wireless Power Compliance between eGaN® FET and MOSFET in a ZVS Class D Amplifier," International Exhibition and Conference for Power Electronics, Intelligent Motion, Renewable Energy and Energy Management (PCIM - Europe), May 2015.

Why eGaN FETs for Wireless Power Transfer?

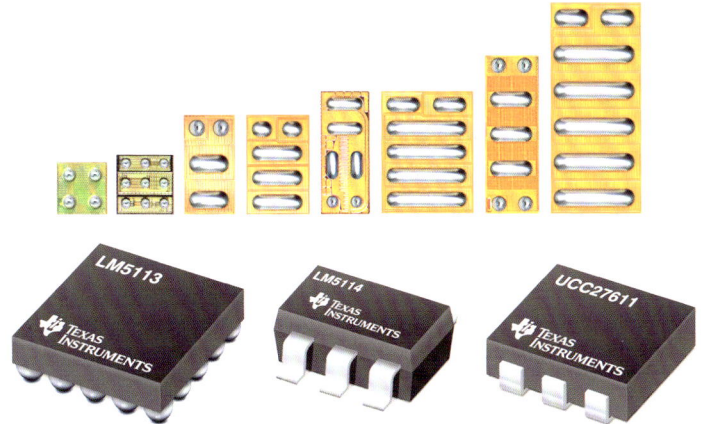

Why eGaN FETs for Wireless Power Transfer?

eGaN FETs make ideal switching devices for use in wireless power transfer due to the following characteristics:

- Low input capacitance (C_{ISS}) and low output capacitance (C_{OSS})
- Zero reverse recovery charge (Q_{RR})
- Low on-state resistance ($R_{DS(on)}$)

These characteristics ensure low operating losses, which lead to higher amplifier efficiency and help keep EMI generation low. In addition, these FETs have a very small footprint and are low profile, which are important for mobile applications.

There are many gate drivers now available that specifically target eGaN FETs. For example, the following are drivers from Texas Instruments:

- LM5113 [4.4] – 100 V capable half-bridge gate driver
- LM5114 [4.5] – single gate driver
- UCC27611 [4.6] – single gate driver with internal voltage regulator

These gate drivers significantly simplify the design as they have been optimized for use with eGaN FETs, thus reducing overall system component count and board area.

eGaN FETs exhibit high dv/dt immunity, which means that regardless of the magnitude of dv/dt across the drain to source, the gate voltage will never exceed the threshold (V_{TH}) and cause unwanted turn-on of the device.

[4.4] Texas Instruments, "5A, 100V Half-Bridge Gate Driver for Enhancement Mode GaN FETs," LM5113 datasheet, June 2011 [Revised April 2013]. [Online] Available: http://www.ti.com/lit/ds/symlink/lm5113.pdf

[4.5] Texas Instruments, "Single 7.6A Peak Current Low-Side Gate Driver," LM5114 datasheet, Jan 2011 [Revised March 2013]. [Online] Available: http://www.ti.com/lit/ds/symlink/lm5114.pdf

[4.6] Texas Instruments, "4A/6A High-Speed 5-V Drive Optimized Single Gate Driver," UCC27611 datasheet, Dec 2012. [Online] Available: http://www.ti.com/lit/ds/symlink/ucc27611.pdf

Discrete eGaN FETs Evaluated in Wireless Power

EPC Part Number	Package mm	V_{DS} (V)	V_{GS} (V)	$R_{DS(on)}$ at 5 V (mΩ)	Q_G at 5 V Typ. (nC)	Q_{GS} Typ. (nC)	Q_{GD} Typ. (nC)	R_G Typ. (Ω)	V_{TH} Typ. (V)	Q_{RR} (nC)	I_D (A)	T_J Max. (°C)
EPC2014C	LGA 1.7 x 1.1	40	6	16	2.0	0.7	0.3	0.4	1.4	0	10	150
EPC2016C	LGA 2.1 x 1.6	100	6	16	3.4	1.1	0.55	0.6	1.4	0	18	150
EPC2007C	LGA 1.7 x 1.1	100	6	30	1.6	0.6	0.3	0.4	1.4	0	6	150
EPC2012C	LGA 1.7 x 0.9	200	6	100	1.0	0.3	0.2	0.6	1.4	0	5	150
EPC2019C	LGA 2.77 x 0.95	200	6	50	1.8	0.6	0.35	0.4	1.4	0	8.5	150
EPC2010C	LGA 3.6 x 1.6	200	6	25	3.7	1.3	0.7	0.4	1.4	0	22	150

Discrete Low Voltage eGaN FETs Evaluated in Wireless Power

This table shows the basic device parameters for some of the low voltage, ultra-low on resistance eGaN FETs offered by EPC. eGaN FETs are available in wafer chip-scale package and Land Grid Array (LGA) with solder bumps on the bottom side of the device. This chip-scale package is possible due to the FET's lateral structure. For the large devices, the source and drain are interleaved to keep inductance low.

Different wireless power transfer topologies and power levels will have different device requirements. As examples, the EPC2012 [4.7] has been successfully tested in class E up to 30 W load power [4.8]–[4.11] and the EPC2014 [4.12] and EPC2007 [4.13] have been successfully tested in both the traditional voltage-mode class D up to 15 W load power [4.14, 4.15] and the ZVS class D up to 37 W load power [4.9, 4.10, 4.16]. The EPC2016 [4.17] has been tested successfully in the current-mode class D up to 38 W load power [4.10].

[4.7] Efficient Power Conversion Corporation, "Enhancement Mode Power Transistor," EPC2012 datasheet, Aug 2011 [Revised Oct 2012]. [Online] Available: http://epc-co.com/epc/documents/datasheets/EPC2012_datasheet.pdf

[4.8] A. Lidow, "How to GaN: eGaN® FETs for High Frequency Wireless Power Transfer," *EEWeb: Power Developer Magazine*, pp. 4–9, January 2013.

[4.9] A. Lidow, M. A. de Rooij, "Performance Evaluation of Enhancement-Mode GaN transistors in Class-D and Class-E Wireless Power Transfer Systems," *Bodo's Power Systems*, May 2014, pp. 56–60.

[4.10] M. A. de Rooij, "eGaN® FET based Wireless Energy Transfer Topology Performance Comparisons," *International Exhibition and Conference for Power Electronics, Intelligent Motion, Renewable Energy and Energy Management (PCIM Europe)*, May 2014, pp. 610–617.

[4.11] M. A. de Rooij, "Performance Evaluation of eGaN® FETs in Low Power High Frequency Class E Wireless Energy Converter," *International Exhibition and Conference for Power Electronics, Intelligent Motion, Renewable Energy and Energy Management (PCIM Asia)*, June 2014, pg 19–26.

[4.12] Efficient Power Conversion Corporation, "Enhancement Mode Power Transistor," EPC2014 datasheet, Aug 2011 [Revised July 2013]. [Online] Available: http://epc-co.com/epc/documents/datasheets/EPC2014_datasheet.pdf

[4.13] Efficient Power Conversion Corporation, "Enhancement Mode Power Transistor," EPC2007 datasheet, Sept 2011 [Revised July 2013]. [Online] Available: http://epc-co.com/epc/Products/eGaNFETs/EPC2007.aspx

[4.14] M. A. de Rooij, J. T. Strydom, "eGaN® FET- Silicon Shoot-Out Vol. 9: Wireless Power Converters," *Power Electronics Technology*, pp. 22–27, July 2012.

[4.15] M. A. De Rooij and J. T. Strydom, "eGaN® FETs in Low Power Wireless Energy Converters," *Electro-Chemical Society transactions on GaN Power Transistors and Converters*, Vol. 50, No. 3, pp. 377–388, October 2012.

[4.16] A. Lidow, "How to GaN: Stable and Efficient ZVS Class D Wireless Energy Transfer at 6.78 MHz," *EEWeb: Pulse Magazine*, Issue 126, pp. 24–31, July 2014.

[4.17] Efficient Power Conversion Corporation, "Enhancement Mode Power Transistor," EPC2016 datasheet, Sept 2013 [Revised Sept 2013]. [Online] Available: http://epc-co.com/epc/documents/datasheets/EPC2016_datasheet.pdf

Ultra-High Frequency eGaN FETs Evaluated in Wireless Power Transfer

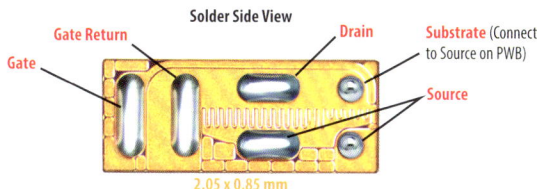

EPC Part Number	Package mm	V_{DS} (V)	V_{GS} (V)	$R_{DS(on)}$ at 5 V (mΩ)	Q_G at 5 V Typ. (pC)	Q_{GS} Typ. (pC)	Q_{GD} Typ. (pC)	R_G Typ. (Ω)	V_{TH} Typ. (V)	Q_{RR} (nC)	I_D (A)	T_J Max. (°C)
EPC8004	LGA 2.05x0.85	40	6	125	358	110	31	0.34	1.4	0	2.7	150
EPC8009	LGA 2.05x0.85	65	6	138	380	116	36	0.3	1.4	0	2.7	150
EPC8010	LGA 2.05x0.85	100	6	160	354	109	32	0.6	1.4	0	2.7	150

Ultra-High Frequency eGaN FETs Evaluated in Wireless Power

This table shows the basic device parameters for the ultra-high frequency eGaN FET products offered by EPC. As with the low voltage, low on-resistance discrete products, these devices are available in Land Grid Array (LGA) chip-scale package with solder bumps on the bottom side of the device.

Due to the very high frequency capability of these devices, a separate gate return has been provided that limits the common source inductance (CSI) to within the device. The Miller ratio of these devices is such that they exhibit full dv/dt immunity [4.18]–[4.20] over the entire rated voltage range. These FETs have their drain and gate circuits orthogonal to each other to further reduce coupling between those circuits.

The EPC8009 [4.21] and EPC8010 [4.22] have been successfully tested in the ZVS class D topology up to 37 W load power [4.23]–[4.25], and the EPC8009 and EPC8004 [4.26] have been implemented in a ZVS class D topology operating at 13.56 MHz [4.24].

[4.18] A. Lidow, J. Strydom, M. de Rooij, D. Reusch, *GaN Transistors for Efficient Power Conversion*. Second Edition, Wiley, ISBN 978-1-118-84476-2, chapter 3.4.

[4.19] A. Lidow, J. Strydom, D. Reusch, "GaN – Moving Quickly into Entirely New Markets," *Power Electronics Europe*, Issue 4, June 2014, pp 28–31.

[4.20] T. Wu, "Cdv/dt induced turn-on in synchronous buck regulators," white paper, International Rectifier Corporation.

[4.21] Efficient Power Conversion Corporation, "Enhancement Mode Power Transistor," EPC8009 datasheet, Sept.. 2013 [Revised Feb. 2015]. [Online] Available: http://epc-co.com/epc/documents/datasheets/EPC8009_datasheet.pdf

[4.22] Efficient Power Conversion Corporation, "Enhancement Mode Power Transistor," EPC8010 datasheet, Dec.. 2013 [Revised Jan. 2015]. [Online] Available: http://epc-co.com/epc/documents/datasheets/EPC8010_datasheet.pdf

[4.23] A. Lidow, "How to GaN: Stable and Efficient ZVS Class D Wireless Energy Transfer at 6.78 MHz," *EEWeb: Pulse Magazine*, Issue 126, pp. 24–31, July 2014.

[4.24] [Online] Available: http://epc-co.com/epc/documents/datasheets/EPC8009_datasheet.pdf

[4.25] [Online] Available: http://epc-co.com/epc/documents/datasheets/EPC8010_datasheet.pdf

[4.26] Efficient Power Conversion Corporation, "Enhancement Mode Power Transistor," EPC8004 datasheet, Sept.. 2013 [Revised Jan. 2015]. [Online] Available: http://epc-co.com/epc/documents/datasheets/EPC8004_datasheet.pdf

Introducing eGaN Integrated Circuit for Wireless Power

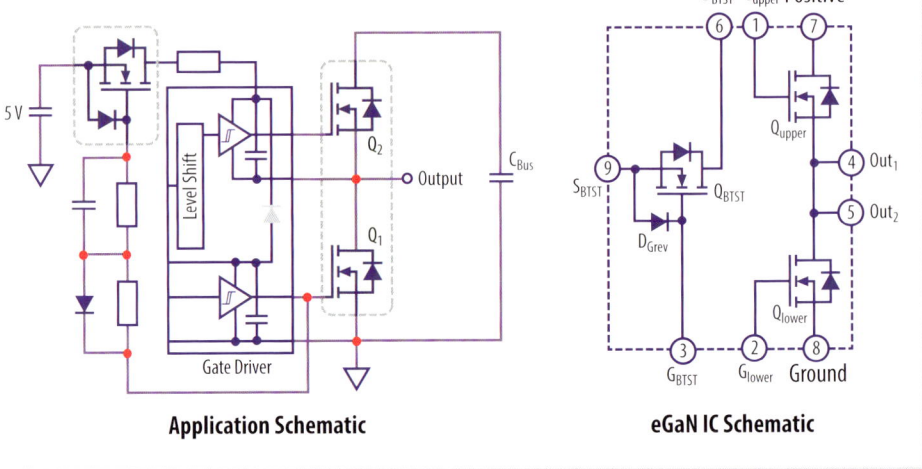

Application Schematic   **eGaN IC Schematic**

Introducing eGaN Integrated Circuit for Wireless Power

With the release of the EPC2100 half-bridge eGaN FET [4.27], EPC has demonstrated the ability for eGaN FETs to be monolithically integrated. Integration not only improves performance but reduces component count and hence cost. In addition, monolithic components simplify layout. These benefits become more apparent in applications requiring higher operating frequencies, such as wireless power transfer.

Furthermore, traditional half-bridge gate drivers make use of a bootstrap diode circuit to provide power to the upper FET [4.28]. This concept works well until the frequency of operation becomes very high as is the case of wireless power. Gate drivers suitable for high frequency operation typically have the bootstrap diode integrated into the gate driver IC, and due to process limitations, this diode will have a significant reverse recovery charge. This reverse recovery charge induces losses in the upper FET, which can become significant at 6.78 MHz. To overcome this high frequency limitation, the bootstrap diode is replaced with an integrated eGaN FET that does not have reverse recovery charge, and implemented in the manner shown above on the left. The benefits of using the synchronous bootstrap FET have been documented in detail [4.29] and was the logical next step to be integrated monolithically with the half-bridge shown above on the right.

[4.27] Efficient Power Conversion Corporation, "Enhancement Mode Power Transistor," EPC2100 datasheet, Aug.. 2014 [Online] Available: http://epc-co.com/epc/documents/datasheets/EPC2100_datasheet.pdf

[4.28] Texas Instruments, "5A, 100V Half-Bridge Gate Driver for Enhancement Mode GaN FETs," LM5113 datasheet, June 2011 [Revised April 2013]. [Online] Available: http://www.ti.com/lit/ds/symlink/lm5113.pdf

[4.29] M. A. de Rooij, "Performance Comparison for A4WP Class-3 Wireless Power Compliance between eGaN® FET and MOSFET in a ZVS Class D Amplifier," International Exhibition and Conference for Power Electronics, Intelligent Motion, Renewable Energy and Energy Management (PCIM - Europe), May 2015.

eGaN FETs and ICs Targeting Wireless Power Applications

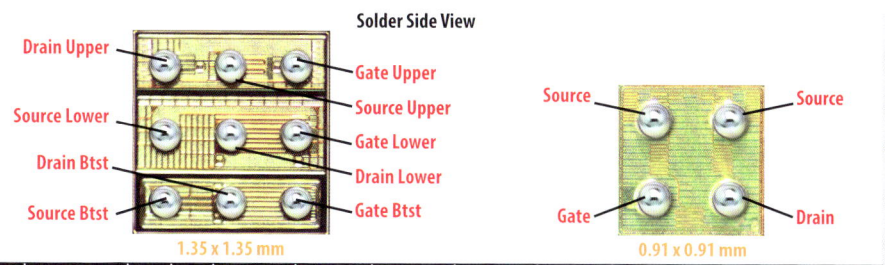

Solder Side View

EPC Part Number	Package mm	V_{DS} (V)	V_{GS} (V)	$R_{DS(on)}$ at 5 V (mΩ)	Q_G at 5 V Typ. (pC)	Q_{GS} Typ. (pC)	Q_{GD} Typ. (pC)	R_G Typ. (Ω)	V_{TH} Typ. (V)	Q_{RR} (nC)	I_D (A)	T_J Max. (°C)
EPC2036	BGA 0.9 x 0.9	100	6	65	700	170	140	0.6	1.4	0	1	150
EPC2037	BGA 0.9 x 0.9	100	6	550	115	30	23	0.6	1.4	0	1	150
EPC2107	BGA 1.35 x 1.35	100	6	240	160	65	40	0.6	1.4	0	1.7	150
	BGA 1.35 x 1.35	100	6	240	160	65	40	0.6	1.4	0	1.7	150
	BGA 1.35 x 1.35	100	6	2800	44	16	5	0.6	1.4	0	0.5	150
EPC2108	BGA 1.35 x 1.35	60	6	150	220	85	45	0.6	1.4	0	1.7	150
	BGA 1.35 x 1.35	60	6	150	220	85	45	0.6	1.4	0	1.7	150
	BGA 1.35 x 1.35	100	6	2800	44	16	5	0.6	1.4	0	0.5	150

eGaN FETs and ICs Targeting Wireless Power Applications

This table shows the basic device parameters for EPC's newest eGaN FETs and integrated circuits that are available in wafer chip-scale package and Ball Grid Array (BGA) with solder bumps on the bottom side of the device. The discrete EPC2036 [4.30] and EPC2037 [4.31] are capable of operating in several wireless power topologies, such as parallel-mode class E [4.32], current-mode class D and ZVS class D.

The benefit of including a synchronous bootstrap FET to the circuit of a half-bridge topology is so beneficial [4.33] that it is included in the 100 V rated EPC2107 [4.34] and 60 V rated EPC2108 [4.35] eGaN half-bridge ICs targeting wireless power applications. The EPC2107 was designed specifically targeting the A4WP class 2 source in a half-bridge ZVS class D topology, and is optimized to yield the highest possible amplifier efficiency. The EPC2108 was designed specifically targeting A4WP the class 3 source in a differential-mode ZVS class D topology and optimized to yield the highest possible amplifier efficiency.

[4.30] Efficient Power Conversion Corporation, "Enhancement Mode Power Transistor," EPC2036 datasheet, Apr. 2015 [Online] Available: http://epc-co.com/epc/documents/datasheets/EPC2036_datasheet.pdf

[4.31] Efficient Power Conversion Corporation, "Enhancement Mode Power Transistor," EPC2037 datasheet, Apr. 2015 [Online] Available: http://epc-co.com/epc/documents/datasheets/EPC2037_datasheet.pdf

[4.32] A. Grebennikov, "Load Network Design Techniques for Class E RF and Microwave Amplifiers," High Frequency Electronics, vol. 3, pp. 18-32, July 2004

[4.33] M. A. de Rooij, "Performance Comparison for A4WP Class-3 Wireless Power Compliance between eGaN® FET and MOSFET in a ZVS Class D Amplifier," International Exhibition and Conference for Power Electronics, Intelligent Motion, Renewable Energy and Energy Management (PCIM - Europe), May 2015.

[4.34] Efficient Power Conversion Corporation, "Enhancement Mode Power Transistor," EPC2107 datasheet, Jul. 2015 [Online] Available: http://epc-co.com/epc/documents/datasheets/EPC2107_datasheet.pdf

[4.35] Efficient Power Conversion Corporation, "Enhancement Mode Power Transistor," EPC2108 datasheet, Jun. 2015 [Online] Available: http://epc-co.com/epc/documents/datasheets/EPC2108_datasheet.pdf

Device Comparison

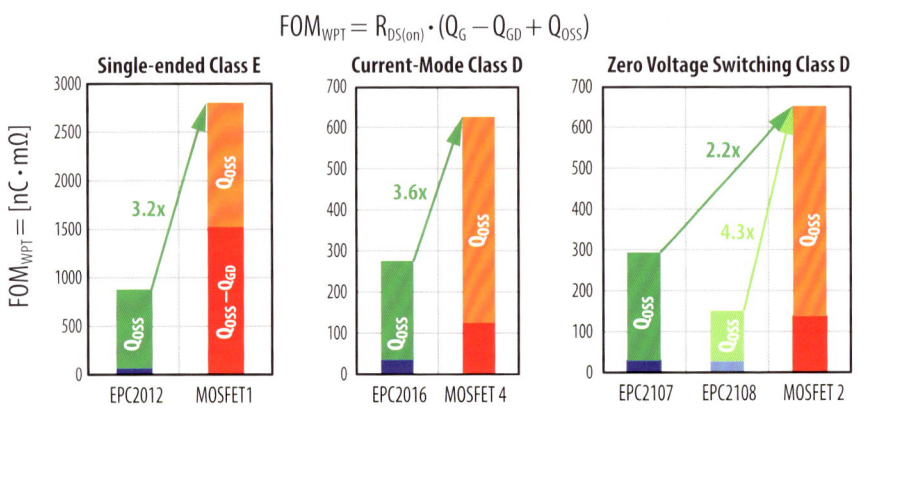

Device Comparison

Using the wireless power FOM_{WPT} for the FETs, we can compare the best-in-class MOSFETs against eGaN FETs in the various amplifier topologies.

First, an eGaN FET (EPC2012) [4.36] will be compared against a FDMC2610 [4.37] in a class E amplifier. Both devices are rated at 200 V and have similar C_{OSS}. Losses due to $R_{\text{DS(on)}}$ are low in comparison to other loss contributors. The $R_{\text{DS(on)}}$ of the eGaN FET is 70 mΩ versus the MOSFET at 175 mΩ. Based on gate charge only, it can be seen that the difference in expected performance is very large at 23.4 times and the eGaN FET barely shows on the comparison graph. Including the Q_{OSS} component, the total FOM_{WPT} is 3.2 times lower than the MOSFET, and indicates that the eGaN FET will yield superior performance in a class E topology.

Next, an EPC2016 [4.38] will be compared against an FDMC86102L [4.39] in a current-mode class D amplifier. Both devices are rated at 100 V and have similar C_{OSS}. The $R_{\text{DS(on)}}$ of the eGaN FET is 12 mΩ versus the MOSFET at 25 mΩ. Based on gate charge only, the expected difference in performance is large at 3.6 times. Including the Q_{OSS} component, the total FOM_{WPT} is 2.3 times lower than the MOSFET, again making the eGaN FET a clear choice.

Device Comparison - *continued*

Lastly, the EPC2107 [4.40] and EPC2108 [4.41] will be evaluated against an FDMC86116LZ [4.42] in a ZVS class D amplifier. Both the EPC2107 and MOSFET devices are 100 V rated and the EPC2108 is 60 V rated. The $R_{DS(on)}$ of the EPC2107 at 240 mΩ is more than double that of the MOSFET at 105 mΩ and the EPC2108 at 150 mΩ is 50% higher. This is due to the eGaN FETs being optimized for wireless power over a wide impedance operating range. In this amplifier, Q_{RR} losses are important when the amplifier is operating off resonance, but since the eGaN FET has zero Q_{RR}, this only serves to further differentiate eGaN FETs from MOSFETs. Even excluding the impact of zero Q_{RR}, the expected performance differential is again large at 4.7 times for the EPC2107, and 5.2 times for the EPC2108, based on gate charge only. Including the Q_{OSS} component, the total FOM_{WPT} is 2.2 times lower for the EPC2107, and 4.3 times lower than the MOSFET, making the eGaN FET the best choice.

[4.36] Efficient Power Conversion Corporation, "Enhancement Mode Power Transistor," EPC2012 datasheet, Aug 2011 [Revised Oct 2012]. [Online] Available: http://epc-co.com/epc/documents/datasheets/EPC2012_datasheet.pdf

[4.37] Fairchild, "200V N-Channel UltraFET Trench MOSFET," FDMC2601 datasheet, Sept 2014 [Rev. C3]. [Online] Available: https://www.fairchildsemi.com/datasheets/FD/FDMC2610.pdf

[4.38] Efficient Power Conversion Corporation, "Enhancement Mode Power Transistor," EPC2016 datasheet, Sept 2013 [Revised Sept 2013]. [Online] Available: http://epc-co.com/epc/documents/datasheets/EPC2016_datasheet.pdf

[4.39] Fairchild, "N-Channel Shielded Gate PowerTrench® MOSFET," FDMC86102L datasheet, June 2014 [Rev. C4]. Fairchild. [Online] Available: https://www.fairchildsemi.com/datasheets/FD/FDMC86102L.pdf

[4.40] Efficient Power Conversion Corporation, "Enhancement Mode Power Transistor," EPC2107 datasheet, Jul. 2015 [Online] Available: http://epc-co.com/epc/documents/datasheets/EPC2107_datasheet.pdf

[4.41] Efficient Power Conversion Corporation, "Enhancement Mode Power Transistor," EPC2108 datasheet, Jun. 2015 [Online] Available: http://epc-co.com/epc/documents/datasheets/EPC2108_datasheet.pdf

[4.42] Fairchild, "N-Channel Shielded Gate PowerTrench® MOSFET," FDMC86116LZ datasheet. Nov 2013 [Rev. C2]. [Online] Available: https://www.fairchildsemi.com/datasheets/FD/FDMC86116LZ.pdf

CHAPTER 5:

Wireless Power Transfer – Experimental Verification

WiTricity® coil set

On-resonance operation

Comparison of amplifier performance

Impact of load variation

Wireless Power Transfer – Experimental Verification

Now that the various amplifier topologies and suitable devices have been presented, the next step will be to evaluate those amplifiers and devices in experimental wireless power transfer systems. Initially this will be done using a WiTricity® designed coil set [5.1]. The system will be tested with the coil set tuned to be on-resonance only (Off-resonance operation will be discussed in Chapter 7.)

Each of the amplifiers will be used to drive the coil set and the overall performance compared. The same coil set will be used in each case to limit variations in performance to the amplifiers only. In addition, the impact of load variation on the performance of the system will be investigated as part of this practical product implementation evaluation.

[5.1] Witricty Corp. coil set part numbers 190-000037-01 and 190-000038-01. [Online] Available: www.witricity.com

Wireless Power Transfer Experimental Setup

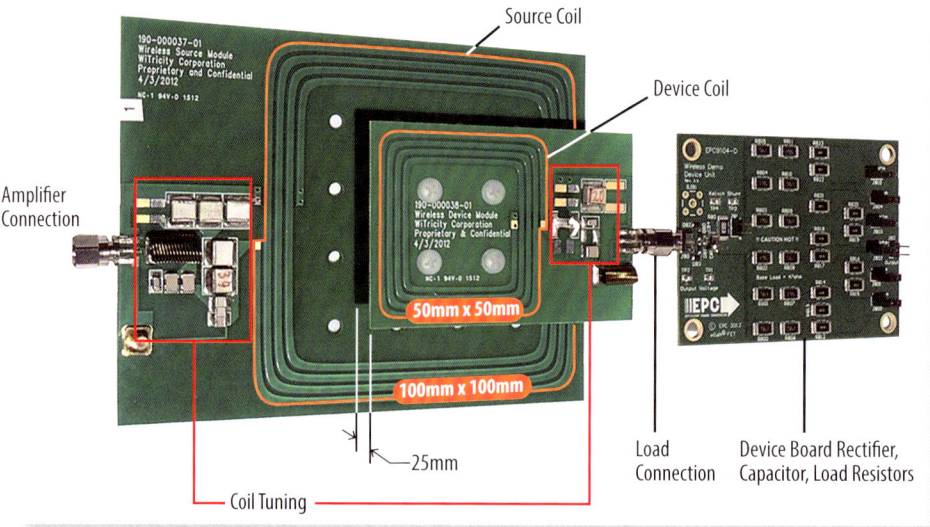

Wireless Power Transfer Experimental Setup

The WiTricity experimental coil set [5.2], together with the load unit is shown here. This coil set has three boards, namely; a source coil (transmit coil) with coil tuning [5.3], a device coil (receive coil) with coil tuning [5.3], and a device board (rectifier and load). The device load board can be pre-set using discrete steps to a DC load resistance range from 14 Ω through 47 Ω. A maximum DC load resistance, which is the lightest load setting, is required. Note that too-high of a load resistance will result in over-voltage, whereas load resistance that is too low will result in over-current and poor energy transfer efficiency. The device employs a full-bridge Schottky diode rectifier for simplicity.

[5.2] Witricty Corp. coil set part numbers 190-000037-01 and 190-000038-01. [Online] Available: www.witricity.com

[5.3] D. M. Pozar, *Microwave Engineering*. Third Edition, J. Wiley, 2005.

On-Resonance Performance Testing Overview

Tests performed:

- Peak Performance – fixed load resistance, variable supply voltage

- Load Variation – fixed supply voltage, variable DC load resistance

- Load Regulation – fixed DC load voltage, variable supply voltage, and variable DC load resistance

Results reported:

- Full System Efficiency – reported from DC input (V_{DD}) to DC output (V_{DCLoad})

On-Resonance Performance Testing Overview

Various tests need to be performed to properly evaluate a wireless power transfer system. These tests target specific characteristics of the wireless energy system and potential application operating conditions.

In this setup, the source and device coils are tuned on-resonance at the design load point, and the tuning remains fixed. Note that the tuning is never re-adjusted during the tests. In these tests, the target DC load resistance is set to 23.6 Ω DC, and the device coil is centrally placed 25 mm above the source coil. The coils must not come closer than four inches from any solid metal surface.

Tests that will be discussed are:

- Peak Performance Testing – A fixed load setting and a variable supply voltage to the amplifier were used. This test measures the optimal performance, or highest potential transfer efficiency of the amplifier and the system. The results of this test are used to compare devices and topologies under ideal operating conditions.

- Load Variation Testing – A fixed supply voltage to the amplifier was used to determine the effect of varying load current on the performance of the system. In such systems, the device could use post regulators to yield a fixed output voltage.

- Load Regulation Testing – With a fixed-load voltage, this test emulates systems designed with output voltage regulation. This test operates the wireless system in the inverse function of its inherent load variation characteristic, making it an aggressive test to perform. Load regulation testing may be required in some cases to meet the severe power dissipation budget limits placed on such systems.

The reported efficiency measurements are for the complete system, from the DC input (V_{DD}) to the DC output (V_{DCLoad}), and includes power consumption from overhead circuits.

Experimental Class E Amplifier

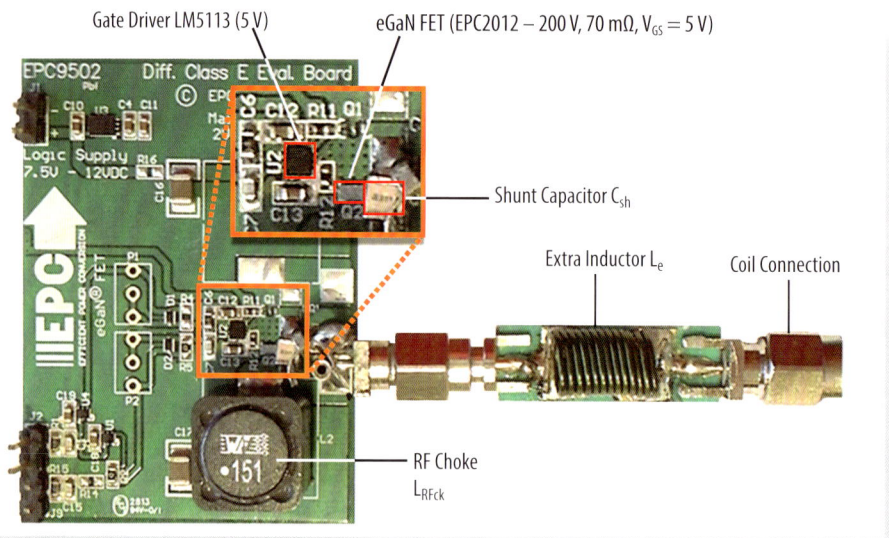

Gate Driver LM5113 (5 V)

eGaN FET (EPC2012 – 200 V, 70 mΩ, V_{GS} = 5 V)

Shunt Capacitor C_{sh}

Extra Inductor L_e

Coil Connection

RF Choke
L_{RFck}

Experimental Class E Amplifier

This photo shows the experimental single-ended class E amplifier that was built using the 200 V rated EPC2012 [5.4] eGaN FETs and LM5113 [5.5] gate driver operating at 5 V. The experimental unit does not have a full controller, but instead is driven by a fixed frequency and duty cycle external oscillator.

The amplifier was designed for optimal operation with a DC load resistance (R_{DCLoad}) of 20.2 Ω. For this design, the external shunt capacitor (C_{sh}) needed was 167 pF and was placed next to the device for minimal inductance in the path between those components. The extra inductor (L_e) required was 390 nH. This extra inductor was placed on a separate external board equipped with a connection to the source coil. The RF choke selected has a large inductance of 150 µH to ensure a stable current source for the amplifier.

[5.4] Efficient Power Conversion Corporation, "Enhancement Mode Power Transistor," EPC2012 datasheet, Aug 2011 [Revised Oct 2012]. [Online] Available: http://epc-co.com/epc/documents/datasheets/EPC2012_datasheet.pdf

[5.5] Texas Instruments, "5A, 100V Half-Bridge Gate Driver for Enhancement Mode GaN FETs," LM5113 datasheet, June 2011 [Revised April 2013]. [Online] Available: http://www.ti.com/lit/ds/symlink/lm5113.pdf

Class E Results – Peak Performance

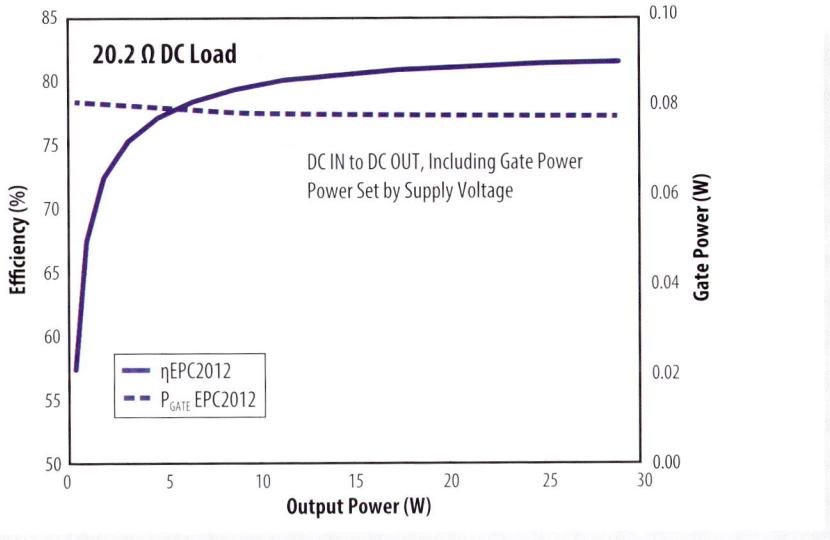

Class E Results – Peak Performance

First, the peak performance efficiency results for the class E amplifier are presented. The results show that under optimal operating conditions, using a DC load resistance (R_{DCLoad}) of 20.2 Ω, the system has a high conversion efficiency that exceeds 80% above 11 W load power.

The efficiency calculation is DC IN to DC OUT. The output power was varied by adjusting the input DC supply voltage (V_{DD}) to the amplifier with a maximum of 28 V and, by doing so, the delivered power reached 29 W. Also included is the gate driver power consumption (P_{gate}), which is around 78 mW.

Class E Results – Load Variation

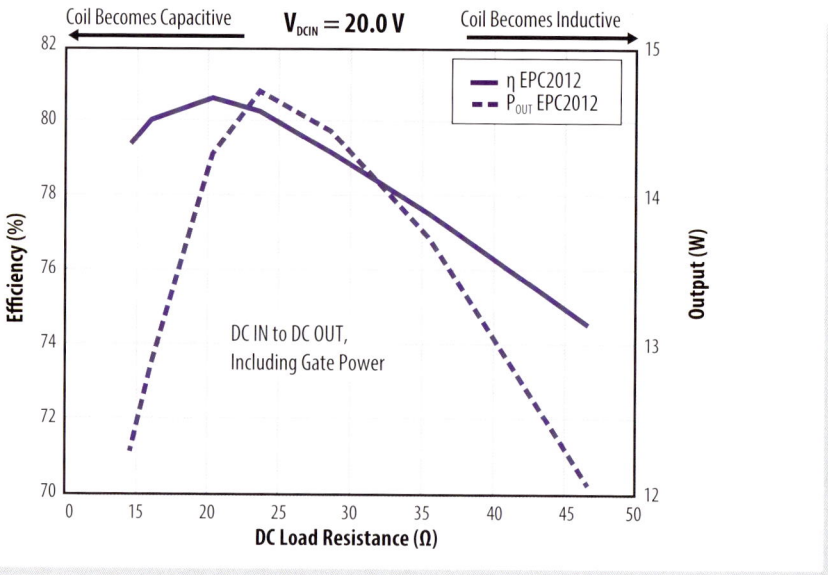

Class E Results – Load Variation

Load variation efficiency results for the class E amplifier are presented. For this test the amplifier was operated with a fixed supply voltage (V_{DD}) of 20 V, and the DC load resistance (R_{DCLoad}) was varied from 14 Ω through 47 Ω in discrete steps. The peak of the curve at a DC load resistance (R_{DCLoad}) of 20.2 Ω shows the design point where the amplifier is operating at its optimal DC load resistance (R_{DCLoad}) point.

The efficiency calculation is DC IN to DC OUT and includes gate driver power consumption. The graph also shows the impact of load variation on the performance of the amplifier, as well as the ability of the system to deliver power that decreases beyond the optimal operating point DC load resistance (R_{DCLoad}) resistance.

Above the optimal design DC load resistance (R_{DCLoad}) point, the amplifier operation is dominated by diode conduction losses due to the reduced energy transition cycle operation of the amplifier. This region is characterized by a linear reduction in system efficiency with increasing DC load resistance (R_{DCLoad}) (decreasing reflected load resistance and the tuned coil impedance becomes more inductive).

Below the optimal design DC load resistance (R_{DCLoad}) point, the amplifier operation is dominated by C_{OSS} losses due to the incomplete energy transition cycle operation. This region is characterized by the exponential reduction in system efficiency with decreasing DC load resistance (increasing reflected load resistance and the tuned coil impedance becomes more capacitive).

Class E Results – Load Regulation

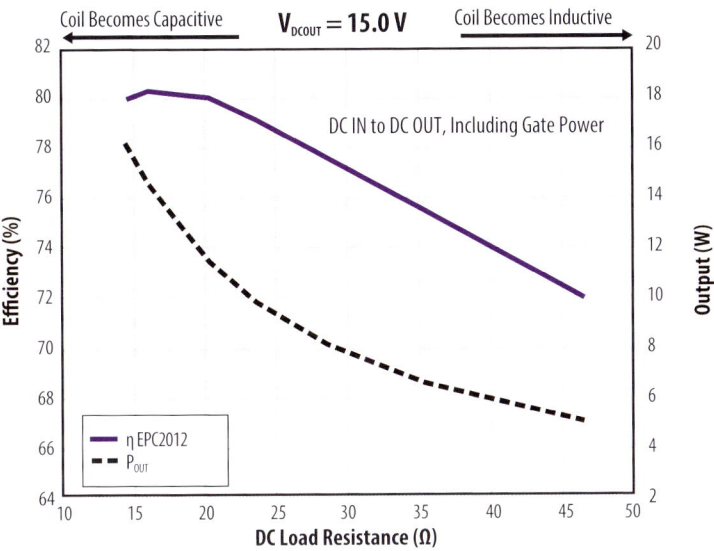

Class E Results – Load Regulation

Load regulation efficiency results for the class E amplifier are presented. For this test the amplifier was operated with a fixed load voltage of 15 V and the supply voltage adjusted according to the DC load resistance (R_{DCLoad}) setting, which was varied from 14 Ω through 47 Ω in discrete steps. The peak at DC load resistance (R_{DCLoad}) of 16 Ω shows that the design point where the amplifier is operating at its optimal point, which has shifted to a lower DC load resistance (R_{DCLoad}) value than the optimal design point. This shift is the result of the output power being forced to operate to a specific point due to the load regulation mode of operation. The indication is that operating the system in load regulation mode is more severe than under load variation conditions. The efficiency calculation is DC IN to DC OUT and includes the gate driver power consumption.

Again, as in the case of load variation tests, above the optimal DC load resistance (R_{DCLoad}) operating point the amplifier operation is dominated by diode conduction losses, which is due to the reduced energy transition cycle operation. This region is characterized by a linear reduction in system efficiency with increasing DC load resistance (R_{DCLoad}), that is, decreasing reflected load resistance (R_{Load}).

In a similar way to the variation tests, below the optimal DC load resistance (R_{DCLoad}) operating point the amplifier operation is dominated by C_{OSS} losses, due to the incomplete energy transition cycle operation. This region is characterized by the exponential reduction in system efficiency with decreasing DC load resistance (R_{DCLoad}) that is increasing reflected load resistance (R_{Load}).

Class E Results – Thermal Performance

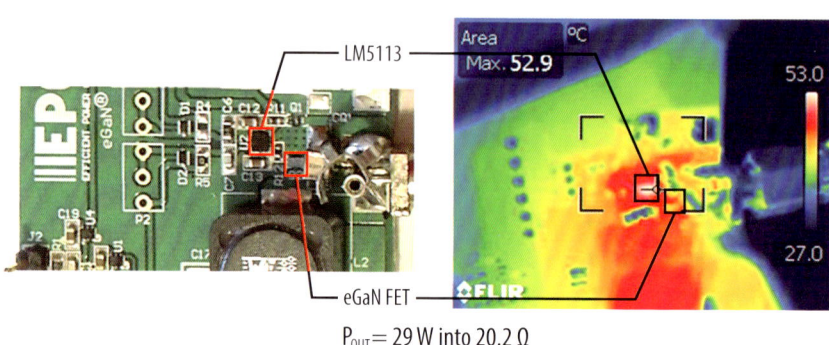

P_{OUT} = 29 W into 20.2 Ω

Class E Results – Thermal Performance

The thermal performance of the class E amplifier operating at maximum load power of 29 W into a DC load resistance (R_{DCLoad}) of 20.2 Ω is shown here. There is no heatsink and no forced air cooling and the circuit was operating in a 25°C ambient. The gate driver is the hottest component on the board yet is well below any commercial temperature limit.

Experimental Current-Mode Class D Amplifier

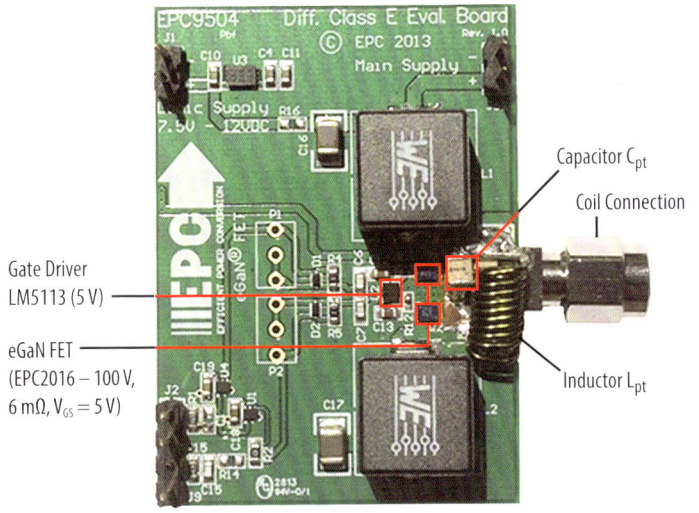

Experimental Current-Mode Class D Amplifier

This photo shows the current-mode class D amplifier that was built using 100 V rated EPC2016 [5.6] eGaN FETs and a LM5113 [5.7] gate driver operating at 5 V. The system does not have a full controller, but instead is driven by a fixed frequency and duty cycle external oscillator. The amplifier was designed for optimal operation with a DC load resistance (R_{DCLoad}) of 23.6 Ω. Also, a parallel tuning capacitance (C_{pt}) of 3 nF, and an inductance (L_{pt}) of 160 nH were used for this design. The choice of RF choke is based on the required ripple specification and minimal impedance impact to the circuit. In this design, a value of 10 µH was chosen for each choke.

[5.6] Efficient Power Conversion Corporation, "Enhancement Mode Power Transistor," EPC2016 datasheet, Sept 2013 [Revised Sept 2013]. [Online] Available: http://epc-co.com/epc/documents/datasheets/EPC2016_datasheet.pdf

[5.7] Texas Instruments, "5A, 100V Half-Bridge Gate Driver for Enhancement Mode GaN FETs," LM5113 datasheet, June 2011 [Revised April 2013]. [Online] Available: http://www.ti.com/lit/ds/symlink/lm5113.pdf

Current-Mode Class D Results – Peak Performance

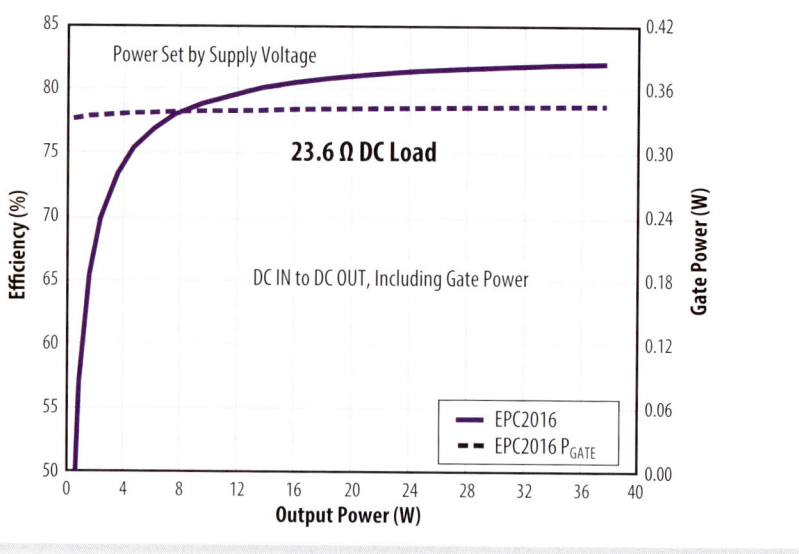

Current-Mode Class D Results – Peak Performance

Peak performance efficiency results for the current-mode class D amplifier are presented. The results show that under optimal operating conditions using a DC load resistance (R_{DCLoad}) of 23.6 Ω, the system exhibits a high conversion efficiency that exceeds 80% above 12 W load power.

The efficiency calculation is DC IN to DC OUT and includes gate driver power consumption. The output power was varied by adjusting the input DC supply voltage (V_{DD}) to the amplifier with a maximum of 10.5 V, and the delivered power reached 38 W. Also included is the gate driver power consumption, which is around 340 mW.

Current-Mode Class D Results – Load Variation

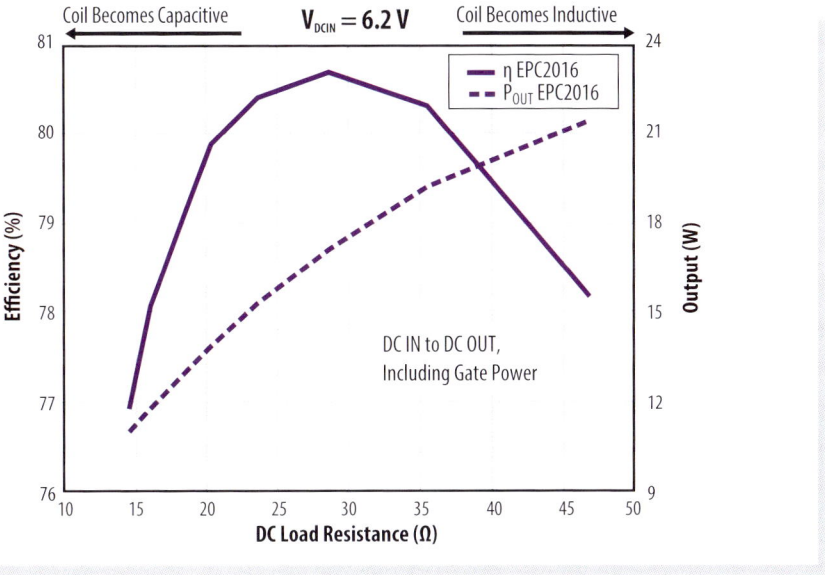

Current-Mode Class D Results – Load Variation

Load variation efficiency results for the current-mode class D amplifier are presented. For this test, the amplifier was operated with a fixed supply voltage of 6.2 V, and the DC load resistance was varied in discrete steps from 14 Ω through 47 Ω. The peak efficiency, located at a DC load resistance (R_{DCLoad}) of 27.5 Ω, shows the point at which the system is operating at its optimal point.

The efficiency calculation is DC IN to DC OUT and includes gate driver power consumption. The graph also shows the impact of load variation on the performance of the amplifier and the ability of the system to deliver power. It is notable that unlike the class E amplifier, the current-mode class D amplifier operates well with the natural characteristic of the coil set where the output power capability improves with increasing DC load resistance (R_{DCLoad}), which is the same as decreasing reflected load resistance (R_{Load}).

Above the optimal DC load resistance (R_{DCLoad}) point, the amplifier operation is dominated by an increase in load inductance, and an associated shift in resonance that causes a decrease in conversion efficiency. Below the optimal point, the amplifier operation is dominated by an increase in load capacitance, and an associated shift in resonance that causes a decrease in conversion efficiency.

Thus, performance of the current-mode class D amplifier is dominated by the performance of the coil set. The output power was only limited by the output voltage capability of the rectifier at 40 V.

Current-Mode Class D Results – Load Regulation

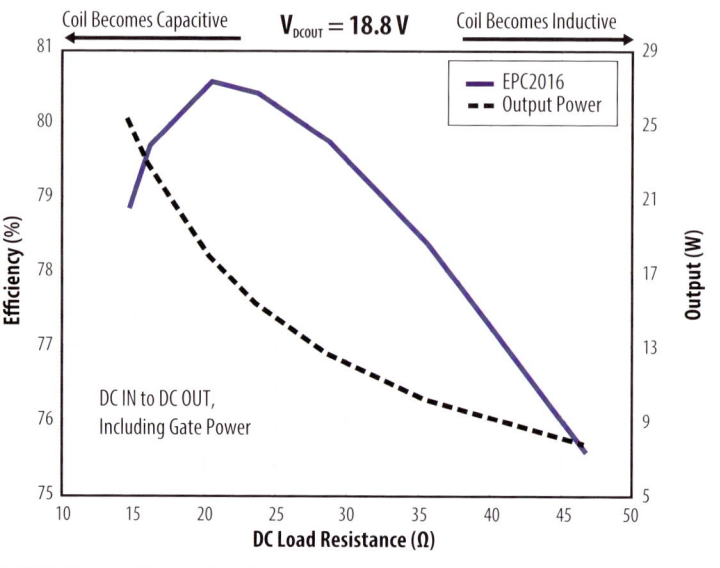

Current-Mode Class D Results – Load Regulation

Load regulation efficiency results for the current-mode class D amplifier is presented. For this test the amplifier was operated with a fixed load voltage of 18.8 V and the supply voltage adjusted according to the DC load resistance (R_{DCLoad}) setting, which was varied from 14 Ω through 47 Ω in discrete steps. The peak, at a DC load resistance (R_{DCLoad}) of 21 Ω, shows the point where the amplifier is operating at its optimal point. It is notable how that point shifted down as a result in the change in operating mode from load variation to load regulation. This is due to the output power being forced to operate at a specific point. The efficiency calculation is DC IN to DC OUT, and includes gate driver power consumption. The graph shows the impact of load regulation on the performance of the amplifier, which again is more severe than under load variation conditions.

As before, in the case of load variation tests above the optimal DC load resistance (R_{DCLoad}) point the amplifier operation is dominated by an increase in load inductance, and an associated shift in resonance that causes a decrease in conversion efficiency. Again, as in the case of load variation tests, below the optimal point the amplifier operation is dominated by an increase in load capacitance and associated shift in resonance that causes a decrease in conversion efficiency.

Current-Mode Class D Results – Thermal Performance

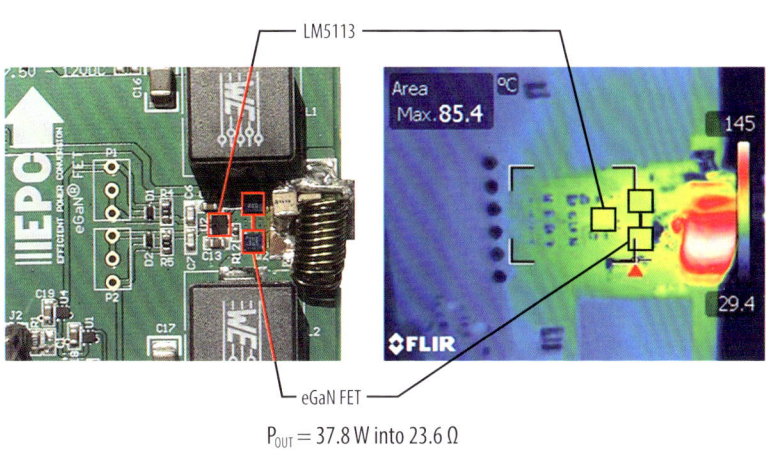

$P_{OUT} = 37.8$ W into 23.6 Ω

Current-Mode Class D Results – Thermal Performance

The thermal performance of the current-mode class D amplifier is presented here when operating at maximum load power of 37.8 W into a DC load resistance (R_{DCLoad}) of 23.6 Ω. There is no heatsink and no forced air cooling and the circuit was operating in an ambient temperature of 25°C. The resonant inductor is the hottest component on the board due to the high magnitude of the current needed to establish resonance.

Experimental ZVS Class D Amplifier

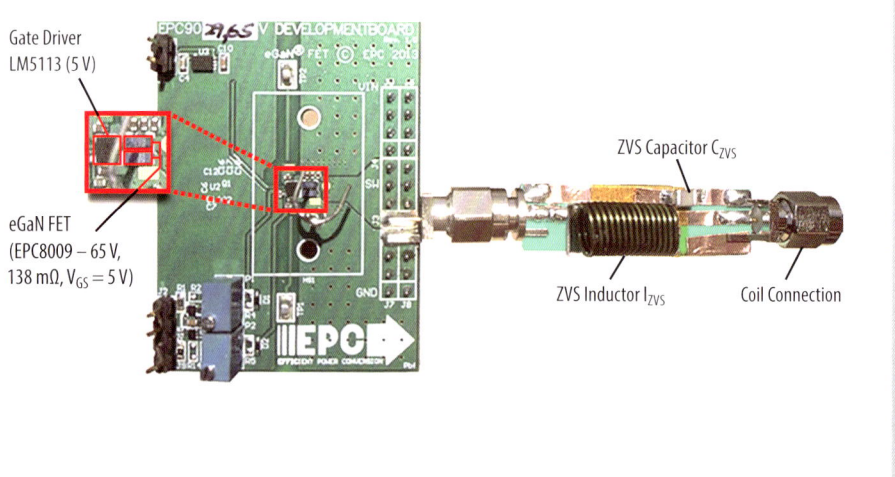

Gate Driver
LM5113 (5 V)

eGaN FET
(EPC8009 – 65 V,
138 mΩ, V_{GS} = 5 V)

ZVS Capacitor C_{ZVS}

ZVS Inductor I_{ZVS}

Coil Connection

Experimental ZVS Class D Amplifier

This photo shows the experimental ZVS class D amplifier that was built using two 65 V EPC8009 [5.8] eGaN FETs, and an LM5113 [5.9] gate driver operating at 5 V using a modified EPC9029 evaluation board [5.10]. The experimental unit does not have a full controller but instead is driven by a fixed frequency and duty cycle external oscillator. The amplifier was designed for optimal operation with a DC load resistance (R_{DCLoad}) of 23.6 Ω. A ZVS inductor of (L_{ZVS}) of 500 nH, and capacitor (C_{ZVS}) of 1 µF were used for this design, both of which were placed on an external board that was connected to the source coil. The external board also served as a connection from the amplifier output to the source coil.

[5.8] Efficient Power Conversion Corporation, "Enhancement Mode Power Transistor," EPC8009 datasheet, Sept 2013 [Revised Sept 2014]. [Online] Available: http://epc-co.com/epc/documents/datasheets/EPC8009_datasheet.pdf

[5.9] Texas Instruments, "5A, 100V Half-Bridge Gate Driver for Enhancement Mode GaN FETs," LM5113 datasheet, June 2011 [Revised April 2013]. [Online] Available: http://www.ti.com/lit/ds/symlink/lm5113.pdf

[5.10] [Online] Available: http://epc-co.com/epc/Products/DemoBoards/EPC9029.aspx

ZVS Class D Results – Peak Performance

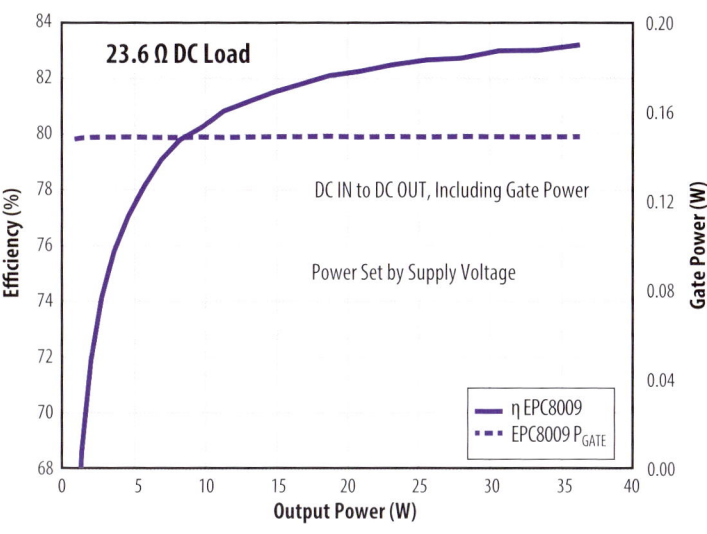

ZVS Class D Results – Peak Performance

Peak performance efficiency results for the ZVS class D amplifier are presented. The results show that under optimal operating conditions using a DC load resistance (R_{DCLoad}) of 23.6 Ω, the system exhibits a high conversion efficiency exceeding 80% above 8.4 W load power. Note that this was the lowest power level to exceed the 80% mark of all the amplifiers tested.

The efficiency calculation is DC IN to DC OUT. The output power was varied by adjusting the input DC supply voltage (V_{DD}), to the amplifier with a maximum of 50 V, and the delivered power reached 36 W. Also included is the gate driver power consumption, which is around 150 mW.

ZVS Class D Results – Load Variation

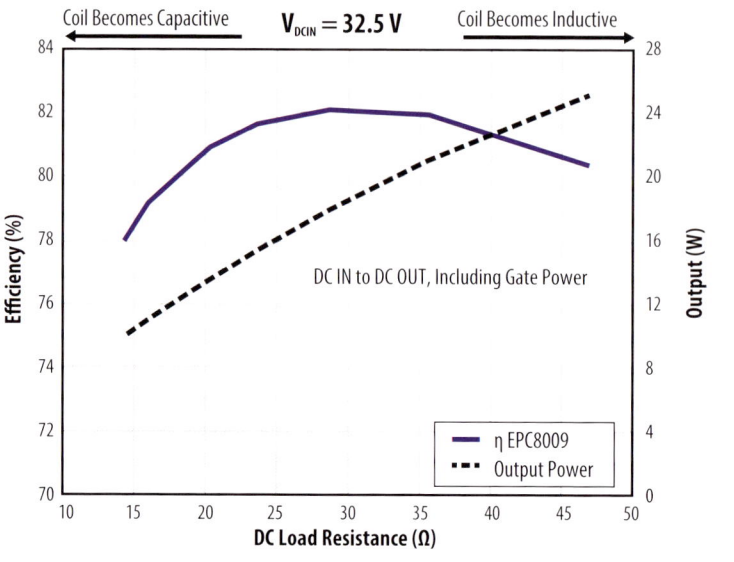

ZVS Class D Results – Load Variation

Load variation efficiency results for the ZVS class D amplifier are presented above. For this test, the amplifier was operated with a fixed supply voltage of 32.5 V and the DC load resistance (R_{DCLoad}) was varied in discrete steps from 14 Ω through 47 Ω. The peak efficiency, located at a DC load resistance (R_{DCLoad}) of 30 Ω, shows the point where the amplifier is operating at its optimal point. The efficiency calculation is DC IN to DC OUT, and includes gate driver power consumption. The graph shows the impact of load variation on the performance of the amplifier and ability of the system to deliver power.

Again, it is notable that in the same manner as the current-mode class D amplifier, the ZVS class D amplifier operates well with the natural characteristic of coil set, where the output power capability improves with increasing DC load resistance (R_{DCLoad}), which is the same as decreasing reflected load resistance (R_{Load}).

Above the optimal DC load resistance (R_{DCLoad}) point, the amplifier operation is dominated by an increase in load inductance and associated shift in resonance that causes a decrease in conversion efficiency.

Below the optimal DC load resistance (R_{DCLoad}) point, the amplifier operation is dominated by an increase in load capacitance and associated shift in resonance that causes a decrease in conversion efficiency.

The performance of the ZVS class D amplifier is dominated by the performance of the coil set.

ZVS Class D Results – Load Regulation

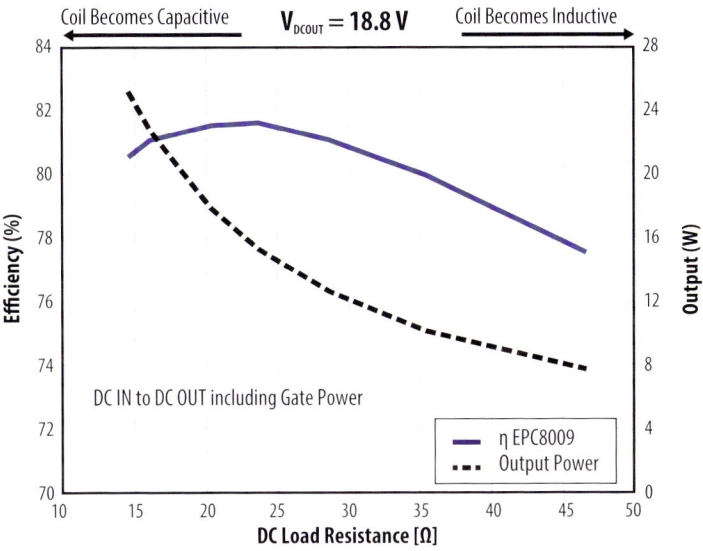

ZVS Class D Results – Load Regulation

Load regulation efficiency results for the ZVS class D amplifier are shown above. For this test, the amplifier was operated with a fixed load voltage of 18.8 V and the supply voltage (V_{DD}) adjusted according to the DC load resistance (R_{DCLoad}) setting. This DC load setting (R_{DCLoad}) was varied in discrete steps from 14 Ω through 47 Ω. The peak at a DC load resistance (R_{DCLoad}) of 23.6 Ω shows the point where the amplifier is operating at its optimal DC load resistance (R_{DCLoad}) point. It is notable how that point shifted down as a result of the change in operating mode from load variation to load regulation. This shift is due to the output power being forced to operate at a specific point, as it did when using the current-mode class D amplifier.

The efficiency calculation is DC IN to DC OUT, and includes gate driver power consumption. The graph also shows the impact of load regulation on the performance of the amplifier, which is more severe than under load variation conditions.

As in the case of load variation tests, above the optimal DC load resistance (R_{DCLoad}) point the amplifier operation is dominated by an increase in load inductance, and associated shift in resonance that causes a decrease in conversion efficiency. Likewise, as in the case of load variation tests, below the optimal DC load resistance (R_{DCLoad}) point the amplifier operation is dominated by an increase in load capacitance, and the associated shift in resonance that causes a decrease in conversion efficiency.

ZVS Class D Results – Thermal Performance

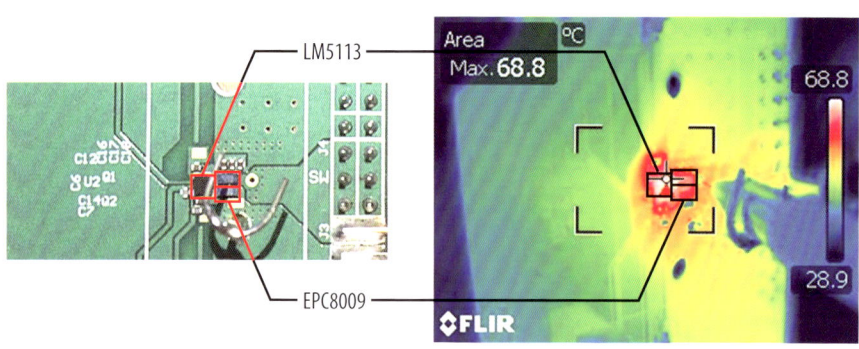

$P_{OUT} = 36$ W into 23.6 Ω

ZVS Class D Results - Thermal Performance

The thermal performance of the ZVS class D amplifier operating at the maximum load power of 36 W into a DC load resistance (R_{DCLoad}) of 23.6 Ω is shown. There was no heatsink and no forced air cooling and the circuit was operating in a 25°C ambient temperature environment. Similar to the class E case, the gate driver is the hottest component, but is still at temperatures below acceptable commercial limits.

Topology Comparison – Peak Performance

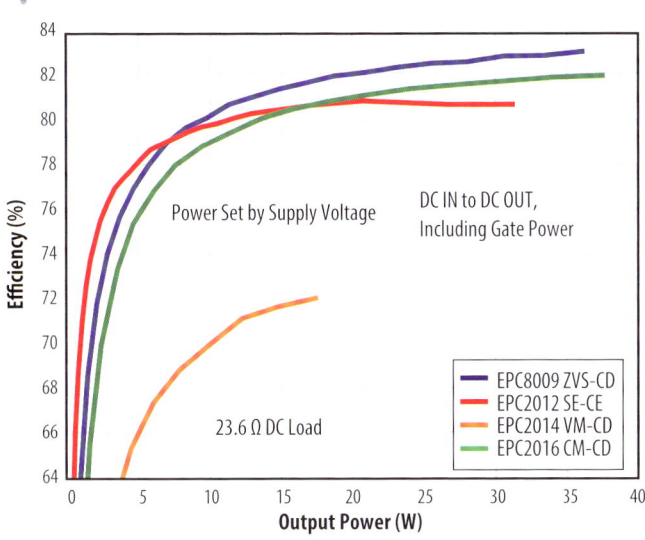

Topology Comparison – Peak Performance

With the various amplifier topologies having been tested under three different conditions, the results can be compared. First, the peak performance efficiency results are presented for each of the three amplifiers tested, and include a fourth amplifier, the traditional voltage-mode class D designed to operate above resonance.

All amplifiers were tested using the same DC load condition. It is clear from the test results that operation above resonance yields a significant reduction in performance when compared to amplifiers that can operate with the load on-resonance. Operation above resonance also reduces the ability of the system to deliver power.

The class E amplifier has higher efficiency in the lower power region (below 7.5 W) due to the lower operating power consumption, primarily due to the low input capacitance (C_{ISS}) of the device. The current-mode class D has the highest operating power for the gate driver, and hence has reduced performance in the low power range (up to 15 W), when compared to the class E or ZVS class D amplifiers. The ZVS class D amplifier has the highest efficiency (above 7.5 W) and maintains that high efficiency performance into the higher power region where the current-mode class D has high losses in the resonant inductor.

Topology Comparison – Load Variation

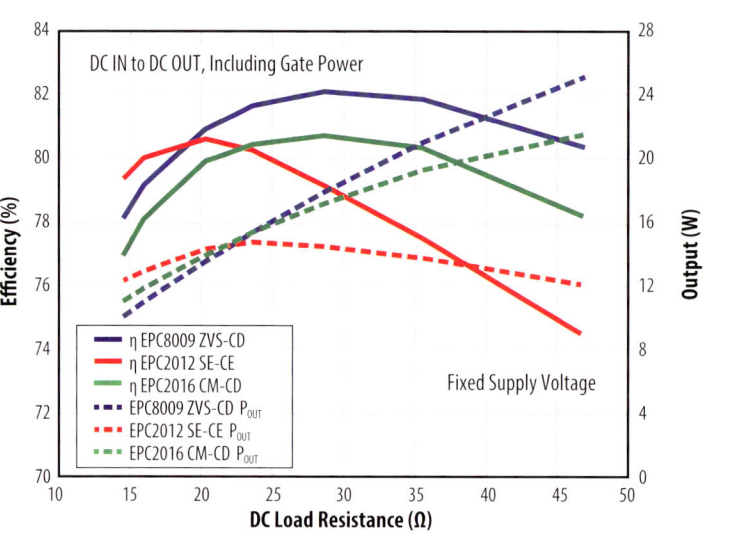

Topology Comparison – Load Variation

Load variation efficiency results for each of the amplifiers are presented. The efficiency calculation is DC IN to DC OUT, and includes gate driver power consumption. The results show that both of the class D type amplifiers have lower performance variation due to DC load resistance (R_{DCLoad}) variation. This is because these amplifiers are less affected by the load variation and the coil set dominates performance. The class E amplifier has a higher variation as a function of DC load resistance (R_{DCLoad}) changes, and also cannot fully make use of the source coil's inherent high power capability in the high DC load resistance range. This is as a result of the class E amplifier performance being most affected by the variation of the coil set.

Topology Comparison – Load Regulation

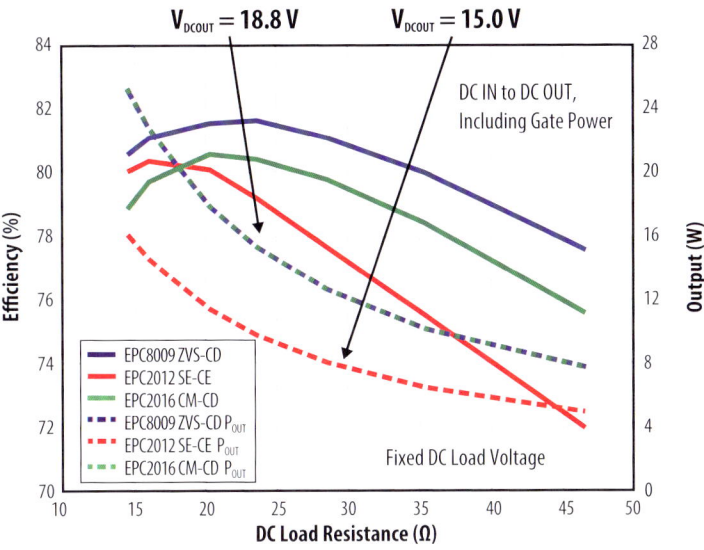

Topology Comparison – Load Regulation

Load regulation efficiency results for of the amplifiers are presented. The efficiency calculation is DC IN to DC OUT, and includes gate driver power consumption. For these tests it is clear that forcing higher power with lower DC load resistance (R_{DCLoad}), as happens with higher reflected resistance, impacts the performance of the system negatively.

In the case of the class E amplifier, it is unable to deliver 18.8 V into the load, unlike the two class D amplifiers. This inability of the class E amplifier to deliver the requisite voltage into the load is due to the thermal limitations of the FET.

CHAPTER 6:

Introducing the Synchronous Bootstrap FET

Gate drivers with internal bootstrap diodes typically have high Q_{RR}

The bootstrap diode Q_{RR} induces losses in the high-side device, where:

- Q_{RR} losses are proportional to frequency
- Such losses are present even with ZVS

Synchronous FET Bootstrap Supply to Circumvent Half-Bridge Gate Driver Limitations

In this section, we look at a method to further improve the efficiency of the ZVS class D amplifier by eliminating the reverse recovery charge (Q_{RR}) losses of the gate driver. Some gate drivers, and the LM5113 [6.1] in particular, are equipped with an internal bootstrap diode [6.2] that is used to provide power to the upper device in a half-bridge configuration. Ideally this diode should be small, have low forward voltage drop, and have no reverse recovery charge – all characteristics of a Schottky diode.

It is very difficult to make a high voltage (100 V) Schottky diode in an IC process and hence the gate drivers with internal bootstrap diode are typically PN junction diodes that have reverse recovery charge. During operation in a half-bridge configuration, this reverse recovery charge of the diode induces losses in the upper transistor. Normally these losses are negligible in comparison to other power loss mechanisms in the circuit. However, the reverse recovery charge losses are proportional to frequency, and at higher operating frequencies will increase to the point where it becomes a significant portion of the converter losses. These losses are present in the ZVS class D amplifier, even though it is operating with ZVS, because the voltage transition time (Δt_{vt}) is shorter than the reverse recovery time (t_{RR}) of the diode. Ways to eliminate these losses are the subject of this chapter.

[6.1] Texas Instruments, "5A, 100V Half-Bridge Gate Driver for Enhancement Mode GaN FETs," LM5113 datasheet, June 2011 [Revised April 2013]. [Online] Available: http://www.ti.com/lit/ds/symlink/lm5113.pdf

[6.2] Fairchild Semiconductor, "Design and Application Guide of Bootstrap Circuit for High-Voltage Gate-Driver IC," Appl. Note AN-6076, Rev. 1.4, December 2014.

Implementation of a Synchronous FET Bootstrap Supply

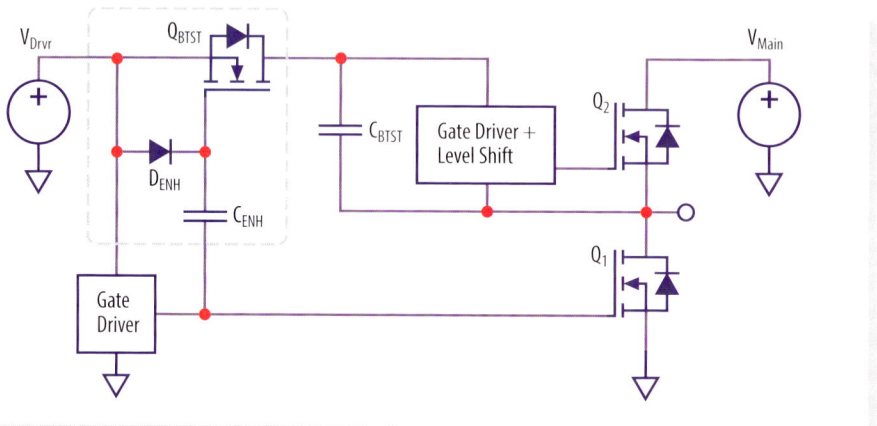

Implementation of a Synchronous FET Bootstrap Supply

Elimination of reverse recovery losses in the ZVS class D amplifier is key to improving efficiency and extending the thermal operating range. The method used shuts down the internal diode of the gate driver and replaces its function by using an external circuit.

Typically, external diodes are large and have reverse recovery too, so this will not work. Instead, an eGaN FET is used to replace the function of the internal bootstrap diode, as they have no reverse recovery charge, are rated to the full voltage of the converter, and have low capacitances (C_{ISS} and C_{OSS}). This implementation only works for eGaN FET technology since MOSFETs have a reverse recovery charge. A unique feature of eGaN FET reverse conduction voltage is that it can be set by a negative gate voltage and results in added margin to prevent unwanted conduction.

The eGaN FET replacing the bootstrap diode will be defined as a synchronous bootstrap FET (Q_{BTST}). To prevent over-voltage of the upper device gate by the lower device "body diode" voltage drop (if it occurs), the synchronous bootstrap FET (Q_{BTST}) will be switched synchronously with the output FET (Q_1). This prevents "body diode" conduction from over charging the upper device bootstrap supply voltage.

Since the eGaN FET is an enhancement-mode device, a small amount of level-shifting is required to drive the gate of the synchronous bootstrap FET. This is achieved using D_{ENH} and C_{ENH} in the circuit shown. When the lower device is held off (0 V), then the diode (D_{ENH}) conducts, and charges C_{ENH} to approximately 5 V (V_{Drvr}). The choice of diode for D_{ENH} can be a Schottky diode rated at 20 V or less, and can be very small so it too has little capacitance to charge and discharge. When the lower device is turned on, its gate voltage will rise to 5 V, the gate voltage of the synchronous bootstrap FET will rise to 10 V, but is only 5 V with respect to its source, and hence it will also turn on. When the lower device is held off, the voltage on the synchronous bootstrap FET gate will be 5 V or 0 V with respect to its source. It is important to note that this implementation does not need an additional active gate driver circuit.

Impact of Gate Driver Q_{RR} on ZVS Class D

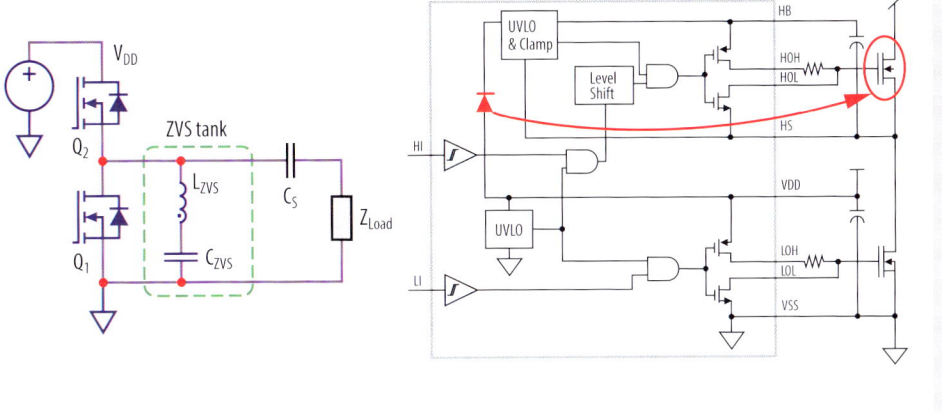

Impact of Gate Driver Q_{RR} on ZVS Class D

An example is used to determine the impact of driver Q_{RR} losses on the ZVS class D amplifier operating at 6.78 MHz, as found in a wireless power transfer application. In this case, the supply voltage (V_{DD}) will be 50 V and no coil is connected as load. This will isolate the operating power for the circuit from the load. An LM5113 gate driver [6.3], with internal diode reverse recovery charge (Q_{RR}) of 2 nC, will be used to drive the EPC8010 eGaN FETs [6.4], and the peak ZVS current (I_{ZVSpk}) will be set to 1.6 A. In this case, the reverse recovery losses (P_{QRR}) will be 370 mW and are induced in the upper device.

[6.3] Texas Instruments, "5A, 100V Half-Bridge Gate Driver for Enhancement Mode GaN FETs," LM5113 datasheet, June 2011 [Revised April 2013]. [Online] Available: http://www.ti.com/lit/ds/symlink/lm5113.pdf

[6.4] Efficient Power Conversion Corporation, "Enhancement Mode Power Transistor," EPC8010 datasheet, Jan 2014 [Revised Jan 2015]. [Online] Available: http://epc-co.com/epc/documents/datasheets/EPC8010_datasheet.pdf

How to Eliminate Gate Driver Induced Q$_{RR}$

The Texas Instruments LM5113's level-shifter and driver can operate at high frequency

The method used to disable the internal bootstrap diode includes:

- Reducing the low-side supply voltage
- Increasing the high-side supply voltage
- Adding the synchronous bootstrap eGaN FET to replace the bootstrap diode function
- Adding protective measures against high side over-voltage

How to Eliminate Gate Driver Induced Q$_{RR}$

Next we need a practical implementation of the synchronous bootstrap FET to the LM5113 gate driver [6.5], and eliminate the reverse recovery charge losses. The method used to shut down the internal bootstrap diode of the LM5113 gate driver begins by slightly reducing the supply voltage to the driver. The supply voltage must remain within the gate driver datasheet specifications.

Next, the upper device bootstrap supply voltage is increased. This is achieved by the use of the synchronous bootstrap FET with its source connected to 5 V, which is now higher than the supply voltage to the IC. The synchronous bootstrap FET is then added together with its own simple low-voltage bootstrap circuit. Lastly, protective measures must be added to further prevent the high-side from over-voltage.

[6.5] Texas Instruments, "5A, 100V Half-Bridge Gate Driver for Enhancement Mode GaN FETs," LM5113 datasheet, June 2011 [Revised April 2013]. [Online] Available: http://www.ti.com/lit/ds/symlink/lm5113.pdf

Synchronous FET Bootstrap Supply for the LM5113

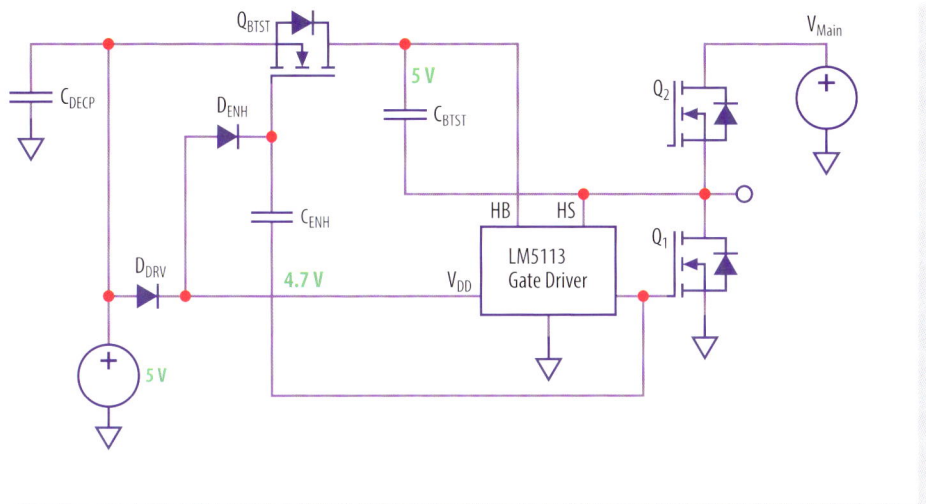

Synchronous FET Bootstrap Supply for the LM5113

This figure shows the practical discrete implementation of the synchronous bootstrap FET to the LM5113 [6.6]. First, D_{DRV} is used to reduce the voltage of the main supply to the gate driver IC by 0.3 V to 4.7 V. D_{ENH} is added to charge C_{ENH} to approximately 4.4 V when the lower device is turned off. A lower voltage is required to hold the synchronous bootstrap FET off with a -0.6 V gate to source, thereby increasing its "body diode" voltage to further prevent it from conducting unexpectedly.

Finally, the synchronous bootstrap FET source is connected to 5 V, and the drain to the HB connection of the gate driver. This results in approximately 5 V across the bootstrap supply capacitor (C_{BTST}). All these measures effectively disable the internal bootstrap diode of the LM5113 gate driver and the synchronous bootstrap FET takes over its function.

[6.6] Texas Instruments, "5A, 100V Half-Bridge Gate Driver for Enhancement Mode GaN FETs," LM5113 datasheet, June 2011 [Revised April 2013]. [Online] Available: http://www.ti.com/lit/ds/symlink/lm5113.pdf

Introducing the Synchronous Bootstrap FET

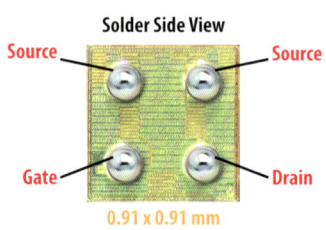

Solder Side View

0.91 x 0.91 mm

EPC Part Number	Package mm	V_{DS} (V)	V_{GS} (V)	$R_{DS(on)}$ at 5 V (mΩ)	Q_G at 5 V Typ. (pC)	Q_{GS} Typ. (pC)	Q_{GD} Typ. (pC)	R_G Typ. (Ω)	V_{TH} Typ. (V)	Q_{RR} (nC)	I_D (A)	T_J Max. (°C)
EPC2038	BGA 0.9 x 0.9	100	6	2800	44	16	5	0.6	1.4	0	0.5	150

Introducing the Synchronous Bootstrap FET

The ability to eliminate Q_{RR} induced losses for high frequency operation is so important that a discrete eGaN FET, EPC2038 [6.7], was specifically designed for this function. This table shows the basic device parameters for this device. It is available in a chip-scale package and Ball Grid Array (BGA) with solder bumps on the bottom side of the device.

Key features of this device are:

- Low C_{OSS}: ~ 1.7 pF
- High $R_{DS(on)}$: ~ 2.8 Ω
- Applicable V_{DS} rating: 100 V

These features minimize the impact of the synchronous FET bootstrap solution on the gate driver and power circuit even when using very small eGaN FETs.

[6.7] Efficient Power Conversion Corporation, "Enhancement Mode Power Transistor," EPC2038 datasheet, Jun 2015 [Online] Available: http://epc-co.com/epc/Portals/0/epc/documents/datasheets/EPC2038_preliminary.pdf

Synchronous Bootstrap FET Design Considerations

Key design requirements for implementation of the synchronous bootstrap FET are:

- **Timing:**
  - Turn on – Delay
  - Turn off – Immediate
- **Off state margin –**
  In case the lower
  FET reverse conducts
- **Drain inductance –**
  Layout critical to prevent ringing

Synchronous Bootstrap FET Design Considerations

To ensure that implementation of the synchronous FET bootstrap circuit has minimal impact to the gate driver and power circuits and operates under any circumstance, it needs to be designed to be robust under the following conditions:

- Low side FET reverse conduction
- Hard switching following low side reverse conduction (a high loss C_{OSS} transition)
- Partial zero voltage switching (PZVS)
- Self-commutation followed by upper FET reverse conduction
- Zero voltage switching (ZVS), the least severe condition

These switching conditions have been well documented in [6.8]. The first key design parameter is related to timing. It takes time for the switch-node voltage to transition from high to low following the rise of the lower FET gate voltage. If the small synchronous FET bootstrap is turned on prematurely, the drain voltage will still be high. Under this condition the switch-node will be connected to the 5 V with disastrous consequences. Therefore, the turn-on of the synchronous FET bootstrap device must be adequately delayed to prevent this from happening. For similar reasons, the turn-off of the synchronous FET bootstrap device must occur at the same time as the main lower FET.

Synchronous Bootstrap FET Design Considerations - *continued*

The next critical design parameter is the off-state margin. Since both the main lower FET and the synchronous FET bootstrap are eGaN FET devices, both will have similar reverse conduction voltage drops when held off. Under these conditions it is possible for the synchronous FET bootstrap device to reverse conduct main current either together or instead of the main lower FET. This can lead to an over-voltage condition across the bootstrap capacitor, which can lead to failure of the upper FET gate. To prevent this from occurring, the off-state voltage of the synchronous FET bootstrap device is held at -1 V, thereby increasing the reverse conduction voltage to be significantly higher than the main lower FET and thus preventing it from conducting.

Finally, the layout of the circuit is critical. In particular, close attention must be paid to the layout of the drain circuit of the synchronous FET bootstrap device, as it can ring to higher voltages that can charge the bootstrap capacitor leading to failure of the upper FET gate.

[6.8]A. Lidow, J. Strydom, M. de Rooij, D. Reusch, "GaN Transistors for Efficient Power Conversion," Second Edition, Wiley, ISBN 978-1-118-84476-2.

Experimental Verification of the Synchronous FET Bootstrap

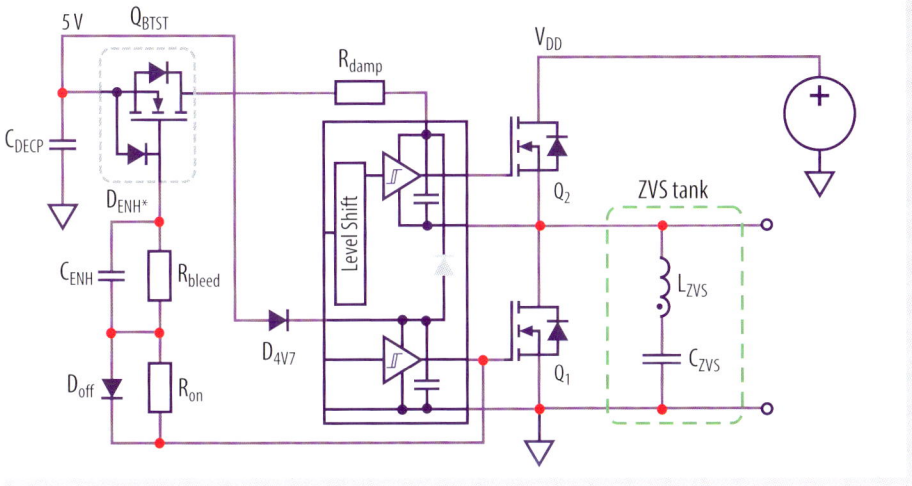

Experimental Verification of the Synchronous FET Bootstrap

The synchronous bootstrap FET concept has been experimentally verified. A ZVS class D amplifier was used for the experimental setup and tested without a load. Two versions were tested; the first at 6.78 MHz using 100 V rated devices and the second at 13.56 MHz also using 100 V rated devices. Each amplifier was provisioned with an enable/disable function for the synchronous bootstrap FET. This allows the same circuit to be used in both modes and the results to be accurately compared.

The first amplifier's main devices (Q_1 & Q_2) are EPC8010 [6.9] (100 V rated, 125 mΩ), and the synchronous bootstrap FET (Q_{BTST}) is the EPC2038 [6.10] (100 V rated, 2800 mΩ). The ZVS tank inductance (L_{ZVS}) is 600 nH, and the capacitor (C_{ZVS}) is 1 μF.

The second amplifier's main devices (Q_1 & Q_2) are EPC8010 [6.9] (100 V rated, 125 mΩ), and the synchronous bootstrap FET (Q_{BTST}) is the EPC2038 [6.10] (100 V rated, 2800 mΩ). The ZVS tank inductance (L_{ZVS}) is 270 nH, and the capacitor (C_{ZVS}) is 1 μF.

[6.9] Efficient Power Conversion Corporation, "Enhancement Mode Power Transistor," EPC8010 datasheet, Jan 2014 [Revised Jan 2015]. [Online] Available: http://epc-co.com/epc/documents/datasheets/EPC8010_datasheet.pdf

[6.10] Efficient Power Conversion Corporation, "Enhancement Mode Power Transistor," EPC2038 datasheet, Jun 2015 [Online] Available: http://epc-co.com/epc/Portals/0/epc/documents/datasheets/EPC2038_preliminary.pdf

Power Dissipation Results

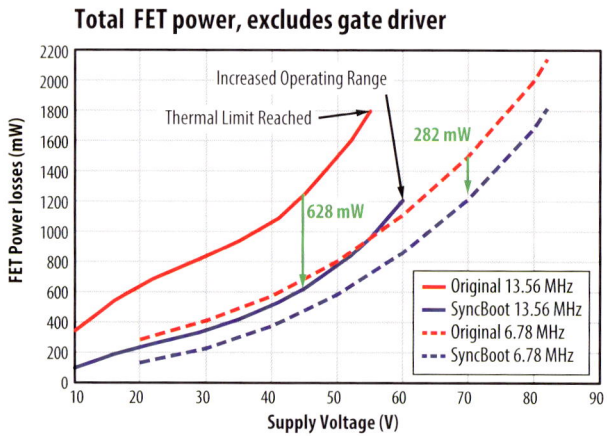

Total FET power, excludes gate driver

Power Dissipation Results

The experimental ZVS class D converter is not driving a load and therefore all power input to the circuit is operating losses. In this setup, losses reported include the power FETs and the ZVS tank circuit, but exclude gate driver and logic circuits.

In the case of the gate driver operating power either includes the reverse recovery losses or the operating losses of the synchronous bootstrap FET circuit. This depends on which circuit is being tested.

For this test, the ZVS class D amplifiers were operated at 6.78 MHz and 13.56 MHz respectively and the higher operating frequency significantly increased the differentiation of the removal of reverse recovery losses with, and without the bypass circuit.

The graphs show the amplifier total power dissipation as function of main supply voltage. The upper red traces shows the original gate driver configuration operating with reverse recovery losses, and the lower blue traces using the synchronous bootstrap FET circuit.

For operation at 6.78 MHz, the ZVS class D amplifier was designed to operate with a supply voltage up to 82 V. The new EPC2038 was used for the synchronous bootstrap FET function. When using the synchronous bootstrap FET at a nominal operating voltage of 70 V, a reduction of 282 mW was achieved.

For operation at 13.56 MHz in the original configuration, the circuit could not be operated above 55 V as the gate driver had reached a thermal limit exceeding 100°C. Using the synchronous bootstrap FET circuit, the main supply voltage could be increased to 60 V, extending the operating range of the amplifier significantly. Most notable was the decrease in power dissipation of 628 mW between the two configurations at the 45 V supply point.

Waveform Improvements at 13.56 MHz

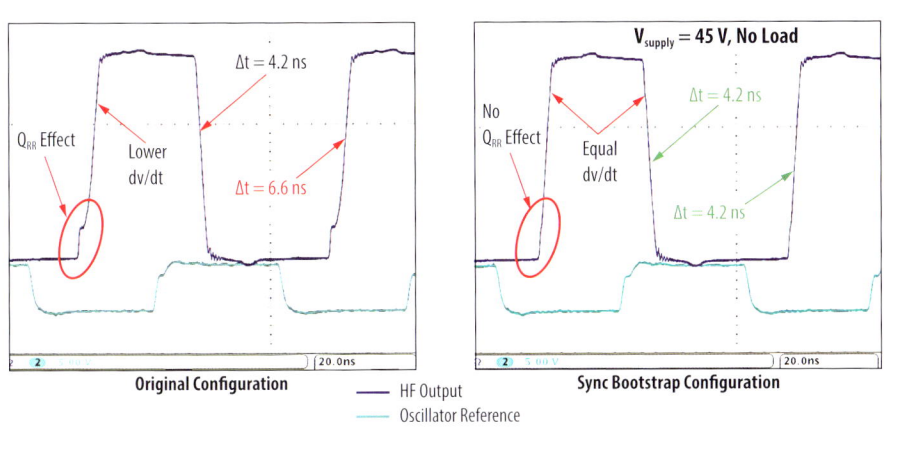

Original Configuration   ——— HF Output   Sync Bootstrap Configuration
                          ——— Oscillator Reference

Waveform Improvements at 13.56 MHz

The measured switch-node waveforms of the ZVS class D amplifier operating at 13.56 MHz is shown in this figure when operating with a supply voltage of 45 V and no load. The left-side dark blue waveform shows the original configuration with reverse recovery charge losses, and the right-side dark blue waveform shows the synchronous bootstrap FET configuration. The light blue waveforms are the oscillator references used to trigger the oscilloscope.

The effect of reverse recovery charge on the left switch-node waveform is visible by looking at the bump near the ground reference that is no longer present on the right waveform. The effect of reverse recovery charge only affects the rising edge of the switch-node waveform. Both (left and right) waveforms' falling edge have the same slope that takes approximately 4.2 ns to complete, and is driven by the magnitude of the supply voltage, output capacitances (C_{OSS}) of the devices, gate driver well capacitance (C_{well}), and magnitude of the ZVS tank circuit current.

When implementing the synchronous bootstrap FET circuit, the rising edge is no longer affected by the gate driver bootstrap diode reverese recovery charge (Q_{RR}), and therefore will now also have the same transition time of 4.2 ns as the falling edge. For this circuit, by eliminating Q_{RR}, the waveform now has excellent slope symmetry.

Thermal Improvements at 6.78 MHz

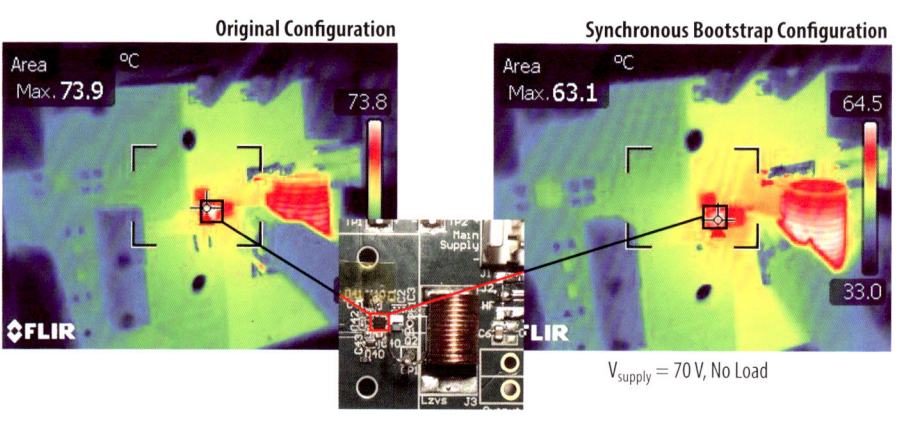

Original Configuration

Synchronous Bootstrap Configuration

$V_{supply} = 70\,V$, No Load

Thermal Improvements at 6.78 MHz

The implementation of the synchronous bootstrap FET circuit, when operating at 6.78 MHz with a main supply of 70 V, decreased the operating power of the circuit by 282 mW. This decrease in power consumption amounts to the decrease in power dissipation of the upper device in the circuit.

Using a thermal camera, it is shown that with Q_{RR} induced losses present, the gate driver temperature is 74°C. After implementation of the synchronous bootstrap FET circuit, the gate driver temperature drops to 63°C, a reduction of 11°C.

Although the gate driver is the hottest component, it is the change in the upper device temperature that reduces the gate driver temperature. This is because the upper FET is placed in close proximity to the gate driver to keep the gate circuit loop inductance down to a minimum and the gate driver has higher thermal resistance to ambient than the eGaN FET.

Thermal Improvements at 13.56 MHz

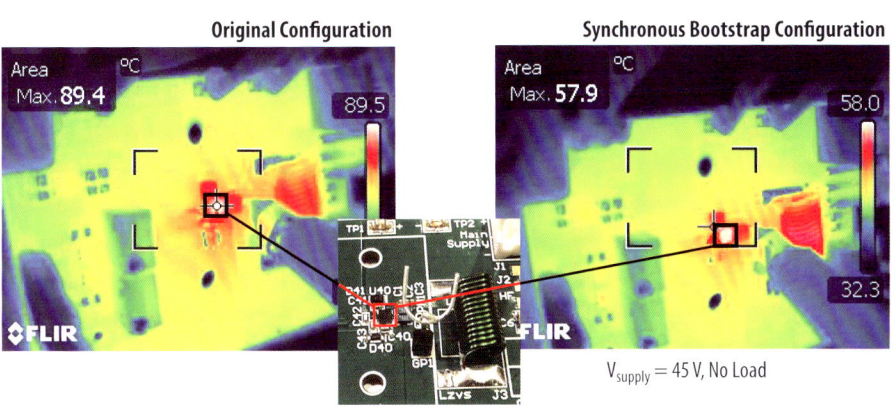

Thermal Improvements at 13.56 MHz

The implementation of the synchronous bootstrap FET circuit, when operating at 13.56 MHz with a main supply of 45 V, decreased the operating power of the circuit by 628 mW. This decrease in power consumption amounts to the decrease in power dissipation of the upper device in the circuit.

Using a thermal camera, it is shown that with Q_{RR} induced losses present the gate driver temperature is 89°C. After implementation of the synchronous bootstrap FET circuit, the gate driver temperature drops to 58°C, a reduction of 31°C.

Although the gate driver is the hottest component, it is the change in the upper device tempera-ture that reduces the gate driver temperature. This is because the upper FET is placed in close proximity to the gate driver to keep the gate circuit loop inductance down to a minimum and the gate driver has higher thermal resistance to ambient than the eGaN FET.

Synchronous FET Bootstrap Supply Summary

The synchronous FET bootstrap supply dramatically improves the ZVS class D efficiency, most notably:

- In low-power systems
- For high-frequency systems

The synchronous FET bootstrap supply is simple to implement and does not require an additional gate driver

Synchronous FET Bootstrap Supply Summary

Wireless power transfer systems are complex and have many cascaded functional blocks. Each block has operating losses that combine and impact overall system efficiency. Any simple means that can be used to dramatically improve efficiency are important. Implementation of a synchronous FET bootstrap supply circuit for the ZVS class D amplifier is just one such approach.

The improvement in efficiency becomes more apparent in lower-power systems where operating power can be on the same order of magnitude as the main throughput power. Amplifiers operating at high frequency will significantly benefit from using the synchronous bootstrap FET approach, as Q_{RR} losses are frequency dependent, with the potential added benefit of increasing the frequency capability of the amplifier. The synchronous bootstrap FET is simple to implement and does not require an additional gate driver, thereby keeping implementation costs down.

CHAPTER 7:

Addressing the Convenience Factor for Wireless Power

Impact of Convenience of Use on the Wireless Coil Set

- Changes in distance between the source and device
- Alignment and position of the device on the source coil
- Multiple devices on a single source
- Introduction of foreign objects (metal or magnetic)

These issues are addressed by the A4WP standard.

Addressing the Convenience Factor for Wireless Power

Up to this point, wireless power transfer using coils operating on-resonance have been presented, analyzed and experimentally verified. Unfortunately, the convenience of use for wireless energy systems is not fully addressed by DC load variations alone because various conditions can significantly detune the coil from its original set point.

De-tuning of the coil set can be caused by variations in distance between the source coil and device, variations in alignment and position of the device on the source coil, placement of multiple devices on the same source coil, or the introduction of a foreign metal object. All these factors will lead to changes in the effective coupling between the coils and tend to shift the reactance, referred to as the imaginary impedance of the coil. The effect of these changes on the performance of the amplifier, addressed by the Alliance for Wireless Power (A4WP) standard [7.1], need to be determined.

[7.1] R. Tseng, B. von Novak, S. Shevde and K. A. Grajski, "Introduction to the Alliance for Wireless Power Loosely-Coupled Wireless Power Transfer System Specification Version 1.0," *IEEE Wireless Power Transfer Conference 2013, Technologies, Systems and Applications*, May 15–16, 2013.

Introducing the A4WP Standard

Source coil – Power transmitting unit (PTU) standards:

- Impedance range of the coil
- Drive requirements for the coil

Source coil requirements drive the amplifier design

Introducing the A4WP Standard

Before delving into the technical aspects of the effects of imaginary impedance variation of the coil set on the performance of the amplifier, the A4WP standard must first be introduced. The discussion will be restricted to the popular class 2 [7.2] and class 3 [7.3] source coils, which are also known as the Power Transmitting Unit (PTU). This standard defines the full impedance range the source coil is expected to present to the amplifier, as well as the drive requirements for the coil. This information will be used to determine its impact on the amplifier, and if additional systems are needed, to fully comply with the standard.

[7.2] A4WP PTU Resonator Class 2 Design - Spiral Type 140-90 A4WP standard document RES-14-0008 RES-14-0006 Ver. 1.2 June 26, 2014.

[7.3] A4WP PTU Resonator Class 3 Design - Spiral Type 210-140 Series (PTU 3-0001), A4WP standard document RES-14-0008 Ver. 1.2, June 30, 2014.

A4WP Class 2 Reflected Impedance Range

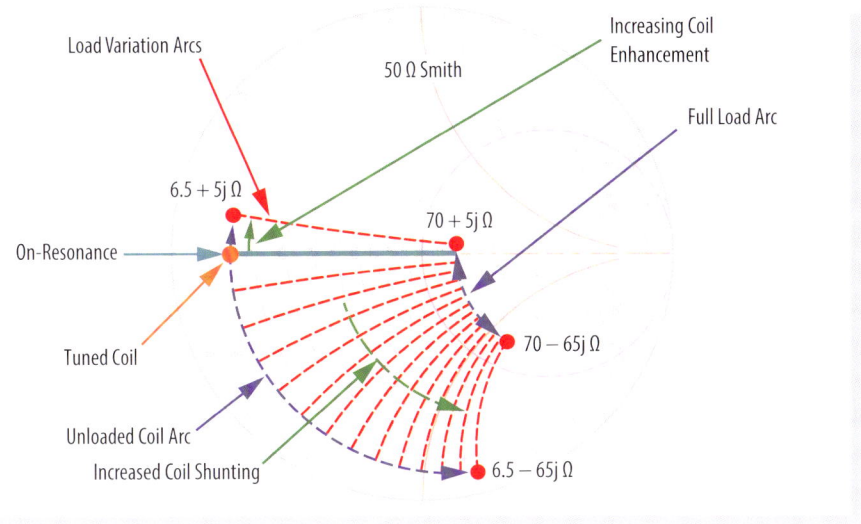

A4WP Class 2 Reflected Impedance Range

The A4WP class 2 source coil (PTU) impedance range [7.4] is represented on a 50 Ω Smith Chart, where the large shaded area is the total tuned compliance region called the "four-corners." This region has been marked on the Smith Chart with the basic values from the standard.

The dashed red arcs represent changes in reflected load resistance caused by changes in load current demand. The lower the DC load current, the further out toward the outer diameter of the graph, the reflected resistance (R_{Load}) will shift, as shown by the outer blue dashed arc arrow.

The higher the DC load current, the closer toward the center of the graph, the reflected load resistance (R_{Load}) will shift, as shown by the inner dark blue dashed arc arrow.

Each of the red dashed arcs represents different reflected imaginary impedances and, the more counter-clockwise the shift, the more negative the imaginary impedance becomes. This is shown by the counter-clockwise green arc arrow and indicates that the load impedance is becoming more capacitive. In the case shown by the clockwise green arc arrow, the load's imaginary impedance is becoming more positive and hence is more inductive.

Negative imaginary impedance shifts are primarily a result of increasing coil shunting, which can be caused by placing a solid metal object on the coil. Positive imaginary impedance shifts that occur above the x-axis are the result of coil enhancement and can occur when a device coil is placed over the source coil under specific conditions.

Finally, the light blue line represents the on-resonance trajectory of the tuned coil impedance. The orange dot is the point where the empty coil was tuned.

[7.4] A4WP PTU Resonator Class 2 Design - Spiral Type 140-90 A4WP standard document RES-14-0008 RES-14-0006 Ver. 1.2 June 26, 2014.

A4WP Class 3 Reflected Impedance Range

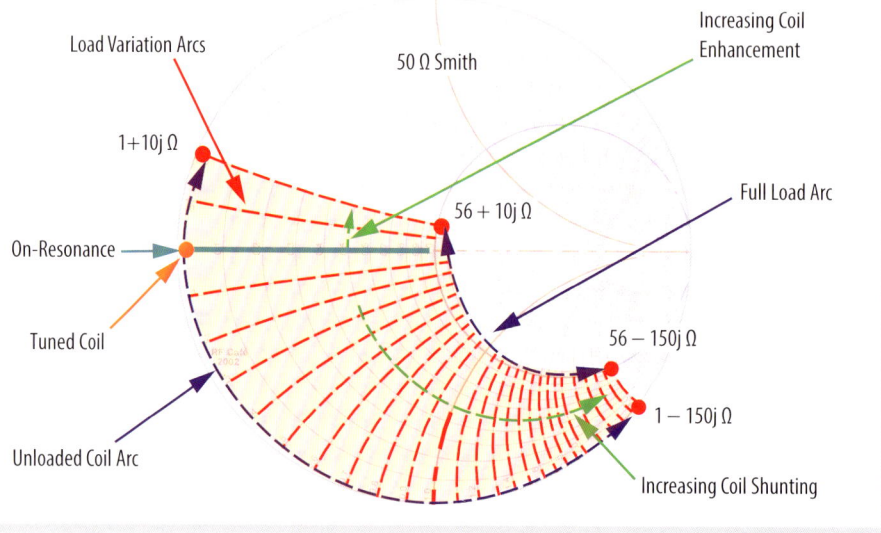

A4WP Class 3 Reflected Impedance Range

The A4WP class 3 source coil (PTU) impedance range [7.5] is represented on a 50 Ω Smith Chart, where the large shaded area is the total tuned compliance region called the "four-corners." This region has been marked on the Smith Chart with the basic values from the standard.

The dashed red arcs represent changes in reflected load resistance caused by changes in load current demand. The lower the DC load current, the further out toward the outer diameter of the graph, the reflected resistance (R_{Load}) will shift, as shown by the outer dark blue dashed arc arrow. The higher the DC load current, the closer toward the center of the graph, the reflected load resistance (R_{Load}) will shift, as shown by the inner blue dashed arc arrow.

Each of the red dashed arcs represents different reflected imaginary impedances and, the more counter-clockwise the shift, the more negative the imaginary impedance becomes. This is shown by the counter-clockwise green arc arrow and indicates that the load impedance is becoming more capacitive. In the case shown by the clockwise green arc arrow, the load's imaginary impedance is becoming more positive and hence is more inductive.

Negative imaginary impedance shifts are primarily a result of increasing coil shunting, which can be caused by placing a solid metal object on the coil. Positive imaginary impedance shifts that occur above the x-axis are the result of coil enhancement and can occur when a device coil is placed over the source coil under specific conditions.

Finally, the light blue line represents the on-resonance trajectory of the tuned coil impedance. The orange dot is the point where the empty coil was tuned.

[7.5] A4WP PTU Resonator Class 3 Design – Spiral Type 210-140 Series (PTU 3-0001), A4WP standard document RES-14-0008 Ver. 1.2, June 30, 2014.

Driving the A4WP Compliant Coil

Class 2 Source Coil drive specifications:
- Nominal coil current = 580 mA$_{RMS}$
- Current is de-rated when P$_{OUT}$ = 10 W

Class 3 Source Coil drive specifications:
- Nominal coil current = 800 mA$_{RMS}$
- Current is de-rated when P$_{OUT}$ = 16 W

Other conditions, but not A4WP requirements:
- Coil tuned with series capacitor (C$_s$) only
- Impedance rotation (on Smith Chart) is permissible to improve efficiency

Driving the A4WP Compliant Coil

The source coil impedance range the amplifier will need to drive for A4WP compliance has been defined and the drive current and power requirements will now be presented.

The A4WP class 2 source coil [7.6] is designated for a maximum output power of 10 W. It is not practical to deliver 10 W across the entire impedance range as this would result in very high current requirements for the amplifier at the lower impedance range. Therefore, the nominal coil current is set to 580 mA$_{RMS}$, where load impedance results in a delivered power of 10 W or less. If 580 mA were to result in output power that would exceed 10 W, then the power should be capped at 10 W with a resultant decrease in coil current.

Similarly, the A4WP class 3 source coil [7.7] is designated for a maximum output power of 16 W, and nominal coil current of 800 mA$_{RMS}$.

In most cases the source coil will be series-tuned to achieve the required current, although this is not an A4WP requirement. However, to cover the wide imaginary impedance range of the A4WP standard it may be necessary to retune the source coil to improve the efficiency of the system. This retuning results in impedance rotation on the Smith Chart and is permissible under the standard.

[7.6] A4WP PTU Resonator Class 2 Design - Spiral Type 140-90 A4WP standard document RES-14-0008 RES-14-0006 Ver. 1.2 June 26, 2014.

[7.7] A4WP PTU Resonator Class 3 Design - Spiral Type 210-140 Series (PTU 3-0001), A4WP standard document RES-14-0008 Ver. 1.2, June 30, 2014.

Implications of the A4WP Reflected Impedance Range

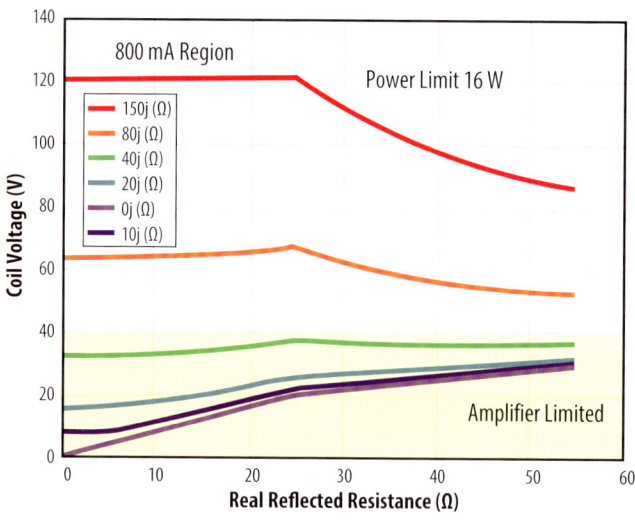

Implications of the A4WP Reflected Impedance Range

We can now analyze the impact of the A4WP class 3 requirements on an amplifier. This analysis will show whether the amplifier is capable of driving the coil set while being fully compliant with the standard on its own, or if additional circuitry is required.

We begin by calculating the voltage range needed to drive the coil if it is only tuned once, resulting in a single tuning set point. This means that the series-tuning capacitor will have only one value. Using the A4WP standard current and power limitations together with the vector sum of the real and imaginary components of the source coil's complex impedance range, the voltage needed to drive the coil can be calculated, and is shown in the graph as a function of the reflected resistance range for various imaginary impedances.

With low magnitude imaginary impedances ($< |40| \text{ j } \Omega$), the voltage needed to drive the source coil remains below 40 V_{RMS}. This is within the capability of most of the amplifiers to operate efficiently (Highlighted by the yellow zone on the graph.) For higher imaginary impedance magnitudes, the range of voltage becomes high, causing the amplifier efficiency to drop at a high rate relative to the increase in the imaginary impedance.

The effects of the imaginary impedance can adversely affect the performance of the amplifier, and is most notable in the case of the class E amplifier, where the imaginary coil impedance will interfere with the amplifier's extra inductance (L_e is connected in series with the coil), and shift the optimal operating point of the amplifier. In addition, the higher the magnitude of the imaginary impedance becomes, the more the coil efficiency will drop. The balance of the voltage required to drive the coil will need to be made up by the amplifier.

Overview of Testing Procedure

The load impedance is emulated using a programmable calibrated load to determine the operating range limits for the amplifier

Operating range limits are set by voltage, current and temperature rating of the active devices in the amplifier

No actual wireless power is transferred during testing

Overview of Testing Procedure

The ability of a wireless power transfer amplifier to drive a source coil in accordance with the A4WP standard must be experimentally tested. The test requires the development of a special load covering the impedance range. This load will need to be calibrated to ensure accuracy, since this load will be used in the amplifier circuit to determine the practical impedance range and power delivered.

The test monitors various operating parameters in the amplifier circuit to determine if they are within predetermined specifications. These specifications are device voltage, temperature, and current limits. For this test, no actual wireless energy is transferred because the calibrated load serves as the reflected impedance range of the coil according to the A4WP standard.

Discrete Programmable Load

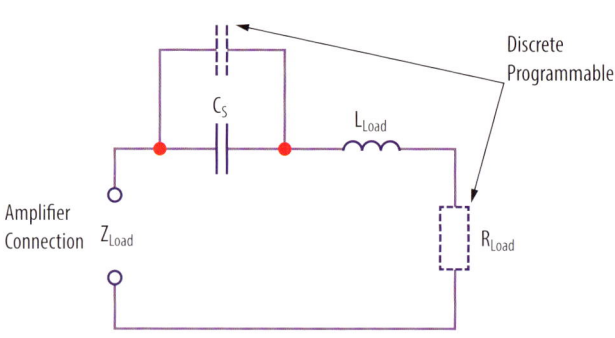

Discrete Programmable Load

The calibrated load that is used for testing is comprised of three main components, namely a resistance (R_{Load}), that can be pre-set to discrete values, a fixed inductance (L_{Load}), and a capacitance (C_S), which also can be pre-set to discrete values. The capacitor is adjustable since it is much easier than changing the inductance value. The values of the capacitors were selected to yield an impedance range from -60j Ω through +45j Ω in steps of 10j Ω or less, if needed. The values of the resistors were selected to yield the following resistance values; 1 Ω, 10 Ω, 14 Ω, 20 Ω, 36 Ω, 44 Ω, and 56 Ω.

In the setup, the current probe used in power testing was included in the calibration, as it contributes both inductance and resistance to the load circuit. The construction of the load had to consider the effects of parasitic elements that can cause unwanted resonance in the circuit and lead to inaccurate measurements. Parasitic capacitance is particularly critical as small values, as low as 25 pF, are easy to establish and are typically of the same order of magnitude to the main tuning capacitances. If the parasitic capacitance is shunting the coil, then this can lead to impedances, low or high, at other unwanted frequencies to become amplified.

Load Calibration Results for Class 2

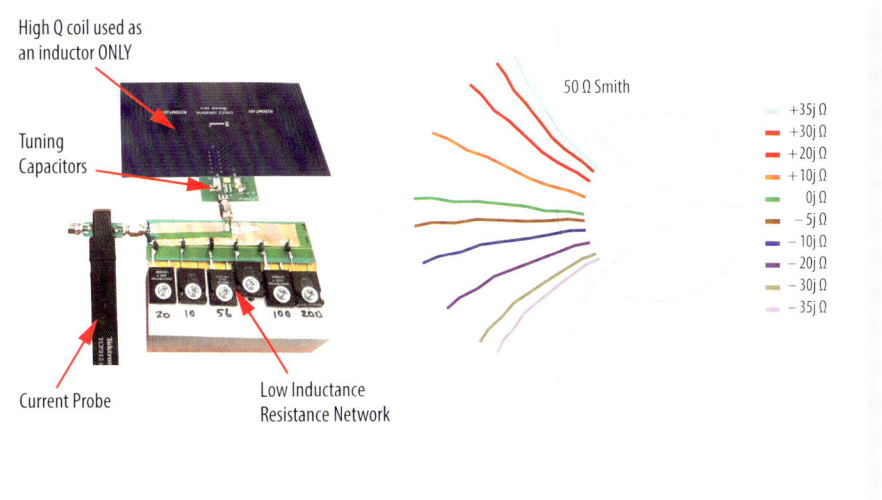

Load Calibration Results for Class 2

A discrete programmable load was constructed and measured using a VNA. This is possible because there are no non-linear components in the load circuit that could lead to measurement errors since the inductor used was air core. The load was measured and calibrated with the oscilloscope current probe attached.

The results of the VNA measurement to the A4WP class 2 standard [7.8] are shown on the Smith Chart as a function of the resistance range and for various imaginary impedances. The measured resistance values can be used to determine the AC power delivered into the load using the current measurement.

[7.8] A4WP PTU Resonator Class 2 Design - Spiral Type 140-90 A4WP standard document RES-14-0008 RES-14-0006 Ver. 1.2 June 26, 2014.

Load Calibration Results for Class 3

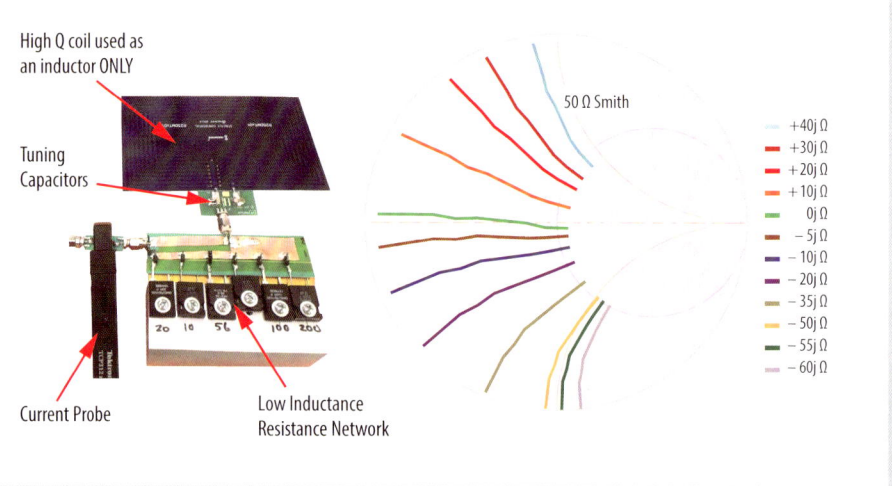

Labels on figure:
- High Q coil used as an inductor ONLY
- Tuning Capacitors
- Current Probe
- Low Inductance Resistance Network
- 50 Ω Smith

Legend:
- +40j Ω
- +30j Ω
- +20j Ω
- +10j Ω
- 0j Ω
- −5j Ω
- −10j Ω
- −20j Ω
- −35j Ω
- −50j Ω
- −55j Ω
- −60j Ω

Load Calibration Results for Class 3

A discrete programmable load was constructed and measured using a VNA. This is possible because there are no non-linear components in the load circuit that could lead to measurement errors since the inductor used was air core. The load was measured and calibrated with the oscilloscope current probe attached.

The results of the VNA measurement to the A4WP class 3 standard [7.9] are shown on the Smith Chart as a function of the resistance range and for various imaginary impedances. The measured resistance values can be used to determine the AC power delivered into the load using the current measurement.

[7.9] A4WP PTU Resonator Class 3 Design - Spiral Type 210-140 Series (PTU 3-0001), A4WP standard document RES-14-0008 Ver. 1.2, June 30, 2014.

A4WP-Based Testing Overview

Amplifiers tested:

- Single-ended Class E to Class 3 standard
- Single-ended ZVS Class D to Class 2 standard
- Single-ended and Differential-Mode ZVS Class D to Class 3 standard

The purpose of testing is:

- Determine maximum impedance range of the amplifier
- Compare MOSFET and eGaN FET performance

A4WP-Based Testing Overview

Armed with the calibrated load and a set of standards to which to measure, the next step is to select amplifiers to test that are capable of efficient wireless power transfer. In this case, the selection has been narrowed to two amplifiers, the class E and ZVS class D. The class E was chosen since it is a popular amplifier with many designers, and the ZVS class D was chosen because it exhibited high efficiency and low variation in performance due to load changes in previous on-resonance only tests.

Various configurations of the amplifiers were constructed that included single-ended class E, single-ended ZVS class D, and differential-mode ZVS class D. The tests are designed to determine the maximum imaginary impedance range that the amplifier can drive and comply with the A4WP standard over the impedance range. Two versions of each amplifier were built for testing and comparison, one using a MOSFET and the other using an eGaN FET. To ensure accurate results, all four amplifiers (class E and ZVS class D used for MOSFET comparison) were tested to a specific imaginary impedance setting before the impedance was changed. The modular structure of the test setup allowed for a quick change of amplifier.

Class 3 Experimental Class E Amplifier

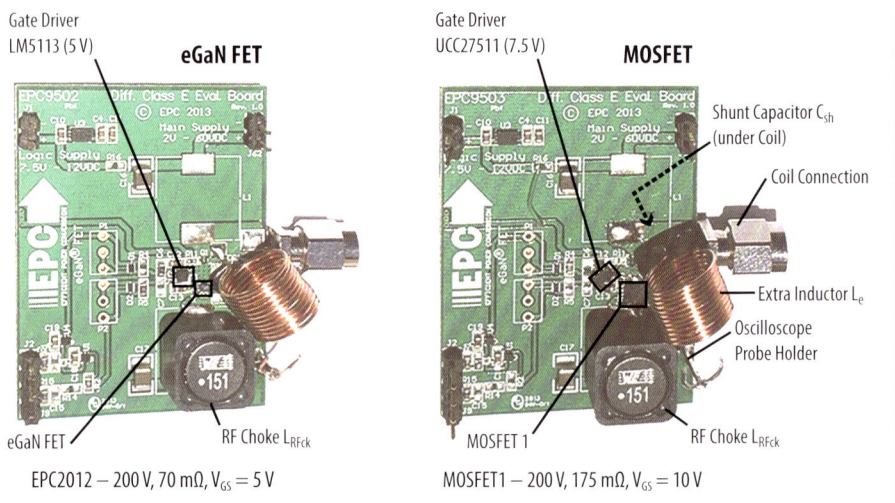

Class 3 Experimental Class E Amplifier

The experimental class E amplifiers are shown in the photo. The EPC eGaN FET (EPC2012) [7.10] version of the amplifier is shown on the left while the Fairchild MOSFET (FDMC2610) [7.11] version is shown on the right.

The eGaN FET version uses the Texas Instruments LM5113 [7.12] gate driver at 5 V and the MOSFET version uses the Texas Instruments UCC27511 [7.13] gate driver at 7.5 V. The optimal design revealed that a 22pF external shunt capacitor was needed for the MOSFET version of the design to correct the C_{OSS} so that both amplifiers would operate to the same specifications. In this version of the amplifier the extra inductor (L_e) was placed on the main board. Both amplifiers used the same 150 µH RF choke.

[7.10] Efficient Power Conversion Corporation, "Enhancement Mode Power Transistor," EPC2012 datasheet, Aug 2011 [Revised Oct 2012]. [Online] Available: http://epc-co.com/epc/documents/datasheets/EPC2012_datasheet.pdf

[7.11] Fairchild, "200V N-Channel UltraFET Trench MOSFET," FDMC2601 datasheet, Sept 2014 [Rev. C3]. [Online] Available: https://www.fairchildsemi.com/datasheets/FD/FDMC2610.pdf

[7.12] Texas Instruments, "5A, 100V Half-Bridge Gate Driver for Enhancement Mode GaN FETs," LM5113 datasheet, June 2011 [Revised April 2013]. [Online] Available: http://www.ti.com/lit/ds/symlink/lm5113.pdf

[7.13] Texas Instruments, "Single-Channel High-Speed Low-Side Gate Driver," UCC27511 datasheet, February 2012 [Revised December 2013]. [Online] Available: http://www.ti.com/product/ucc27511

Class 3 Experimental ZVS Class D Amplifier

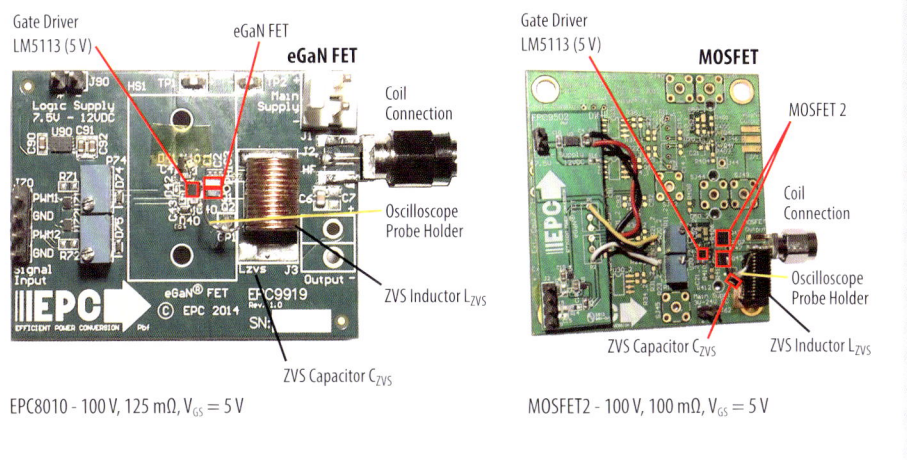

EPC8010 - 100 V, 125 mΩ, V_{GS} = 5 V

MOSFET2 - 100 V, 100 mΩ, V_{GS} = 5 V

Class 3 Experimental ZVS Class D Amplifier

The experimental ZVS class D amplifiers are shown in the photo. The eGaN FET version used a custom designed evaluation board fitted with the 100 V rated EPC8010 [7.14] and included the EPC8010 [7.14] as a synchronous FET bootstrap is shown on the left. The MOSFET (FDMC86116LZ) version is shown on the right. Both the eGaN FET and MOSFET versions use the LM5113 gate driver at 5 V. The design of the ZVS tank inductor (L_{ZVS}) was chosen to yield equal transition times of the switch-node. Since the MOSFET has a higher C_{OSS} than the eGaN FET, a higher current is required to yield the same transition time of approximately 4 ns. The eGaN FET amplifier uses a 600 nH inductor for L_{ZVS} and the MOSFET version uses 390nH.

[7.14] Efficient Power Conversion Corporation, "Enhancement Mode Power Transistor," EPC8010 datasheet, Dec.. 2013 [Revised Jan. 2015]. [Online] Available: http://epc-co.com/epc/documents/datasheets/EPC8010_datasheet.pdf

Experimental Differential-Mode ZVS Class D Amplifier

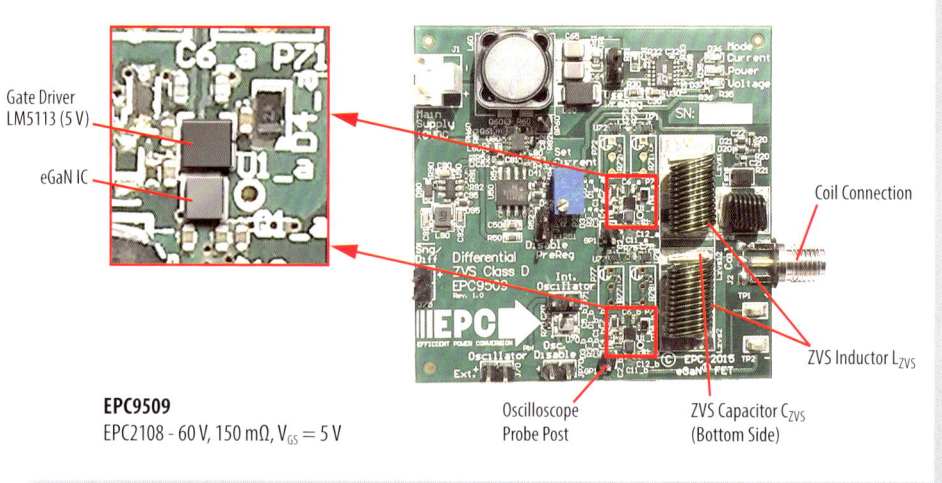

Gate Driver
LM5113 (5 V)

eGaN IC

Coil Connection

ZVS Inductor L$_{ZVS}$

EPC9509
EPC2108 - 60 V, 150 mΩ, V$_{GS}$ = 5 V

Oscilloscope
Probe Post

ZVS Capacitor C$_{ZVS}$
(Bottom Side)

Experimental Differential-Mode ZVS Class D Amplifier

The experimental differential-mode ZVS class D amplifier [7.15] that uses the new 60 V rated EPC2108 [7.16] eGaN integrated circuit is shown in the photo. Although this demonstration board features a controller to operate to the A4WP standard, the controller was bypassed to allow specific measurements to be taken. This amplifier will not be compared to an equivalent MOSFET version, but it will be used to determine the full reactive impedance range drive capability to the class 3 standard. This board also includes a new high frequency current sense circuit which will remain in the circuit. This current sense circuit will shift the reactive impedance range by approximately +5j Ω.

Once the amplifier has been tested it will be used for system level testing using a coil set to deliver power wirelessly.

[7.15] [Online] Available: http://epc-co.com/epc/Products/DemoBoards/EPC9509.aspx

[7.16] Efficient Power Conversion Corporation, "Enhancement Mode Power Transistor," EPC2108 datasheet, Jun. 2015 [Online] Available: http://epc-co.com/epc/documents/datasheets/EPC2108_datasheet.pdf

Experimental Single-Ended ZVS Class D Amplifier

Gate Driver
LM5113 (5 V)

eGaN IC

ZVS Inductor L_{ZVS}

EPC9510
EPC2107 - 100 V, 220 mΩ, V_{GS} = 5 V

Oscilloscope
Probe Post

ZVS Capacitor C_{ZVS}
(Bottom Side)

Coil Connection

Experimental Single-Ended ZVS Class D Amplifier

The experimental single-ended ZVS class D amplifier [7.17] that uses the new 100 V rated EPC2107 [7.18] eGaN integrated circuit is shown in the photo. Although this demonstration board features a controller to operate to the A4WP standard, the controller was bypassed to allow specific measurements to be taken. This amplifier will not be compared to an equivalent MOSFET version but will be used to determine the full reactive impedance range drive capability to the class 2 standard. This board also includes a new high frequency current sense circuit which will remain in the circuit. This current sense circuit will shift the reactive impedance range by approximately +5j Ω.

Once the amplifier has been tested it will be used for system level testing using a coil set to deliver power wirelessly.

[7.17] [Online] Available: http://epc-co.com/epc/Products/DemoBoards/EPC9510.aspx

[7.18] Efficient Power Conversion Corporation, "Enhancement Mode Power Transistor," EPC2107 datasheet, Jul. 2015 [Online] Available: http://epc-co.com/epc/documents/datasheets/EPC2107_datasheet.pdf

Experimental Details

Fail criteria for A4WP compliance testing

- V_{DS} exceeds 80% of rated datasheet value
- T_{FET} or T_{Driver} exceeds 100°C in an ambient temperature of 25°C

All the amplifiers are tested using the same load setting prior to the load being changed

Gate driver power consumption is measured

Efficiency is from DC input to AC power delivered and includes gate driver power

Experimental Details

The experimental tests over a wide impedance range will push each of the amplifiers to their limits. It is not the intention of these tests to find the breaking point of the amplifiers but rather to determine the maximum operating point within well-known guidelines. These guideline limits are set by two main criteria, and are based on commercial product limits and reliability considerations.

In this case, if the operating voltage of the circuit leads to any device to be exposed above 80% of its rated maximum V_{DS} specification, then that would be considered out of bounds. Likewise, if any component exceeds 100°C in an operating ambient temperature of 25°C, it would also be considered out of bounds. Upon detection of an out of bounds condition, that specific measurement will be halted. A determination will be made if the next lower magnitude impedance measurement is needed and if a new test needs to be executed.

All four amplifiers (class E and ZVS class D used for MOSFET comparison) will be tested prior to an imaginary impedance setting change in order to prevent variations in imaginary impedance.

The efficiency reported from these tests will be from the DC input to AC power delivered, and includes the gate driver power consumption.

Class E Amplifier Class 3 Experimental Results

Total Amplifier Losses

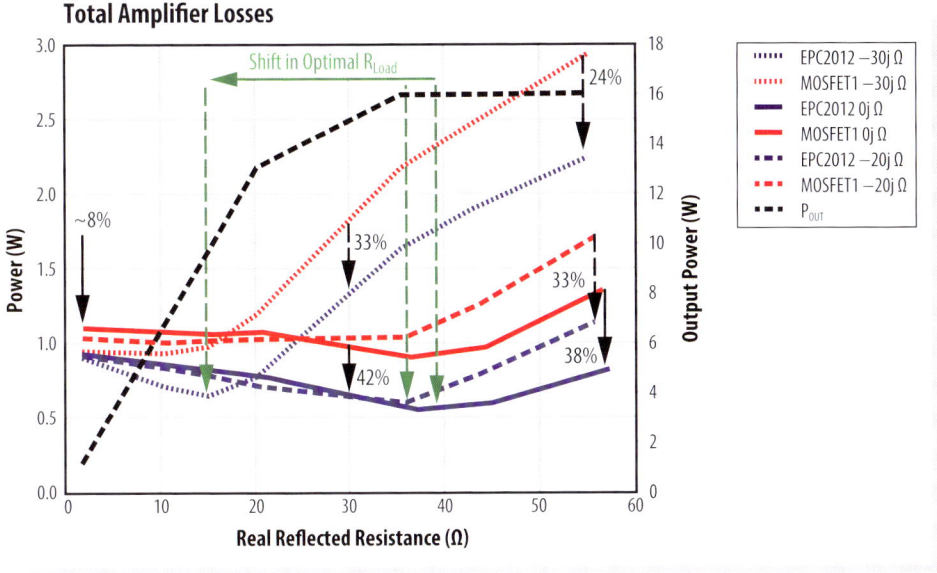

Class E Amplifier Class 3 Experimental Results

Shown above is the measured total amplifier power loss for both the eGaN FET (blue traces) and MOSFET (red traces) based amplifiers operating under a resistance load variation of 1.7 Ω through 57 Ω, and for three imaginary impedance load conditions, where both amplifiers were fully A4WP class 3 compliant, of -30j Ω, -20j Ω, and 0j Ω.

The results reveal that the eGaN FET based amplifier always has lower operating power losses. Furthermore, in the power range from 8 W through 16 W, this amplifier has between 24% and 42% lower losses than the MOSFET based amplifier. Also notable is the shift downward in optimal load resistance as the imaginary impedance decreases (becomes more negative). This is due to the load effectively reducing the value of the extra inductor (L_e) and therefore the optimal operating point resistance.

Single-Ended ZVS Class D Amplifier Class 3 Experimental Results

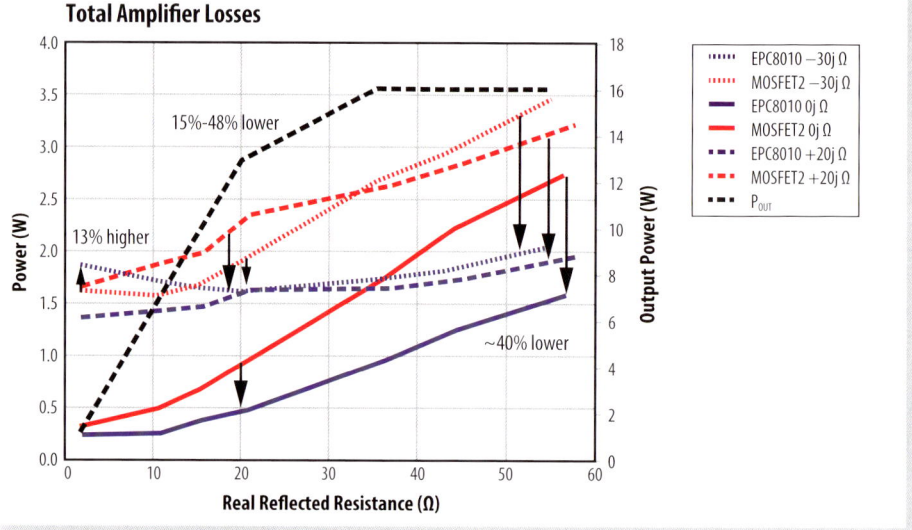

Single-Ended ZVS Class D Amplifier Class 3 Experimental Results

Shown above is the measured total amplifier power loss for both the eGaN FET and MOSFET based amplifiers operating under a resistance load variation of 1.7 Ω through 57 Ω and for three imaginary impedance load conditions, where both amplifiers were fully A4WP class 3 compliant, of -30j Ω, 0j Ω, and +20j Ω.

The results reveal that for the load power range between 6.5 W and 16 W, the eGaN FET amplifier has between 15% and 48% lower losses than the MOSFET version. In the power range below 6.5 W, the eGaN FET based amplifier maintains lower losses, except when the load becomes capacitive. In this case, the difference is small at 13% (in the worst case), and is due to the lack of availability at the time these tests were conducted of a very small synchronous FET bootstrap device with negligible C_{OSS} with respect to the main devices (Q_1 and Q_2). This performance degradation can now be eliminated by using the EPC2038 [7.19] eGaN FET as the synchronous FET bootstrap device.

[7.19] Efficient Power Conversion Corporation, "Enhancement Mode Power Transistor," EPC2038 datasheet, Jun 2015 [Online] Available: http://epc-co.com/epc/Portals/0/epc/documents/datasheets/EPC2038_preliminary.pdf

Comparison Between Single-Ended ZVS Class D and Class E Amplifiers

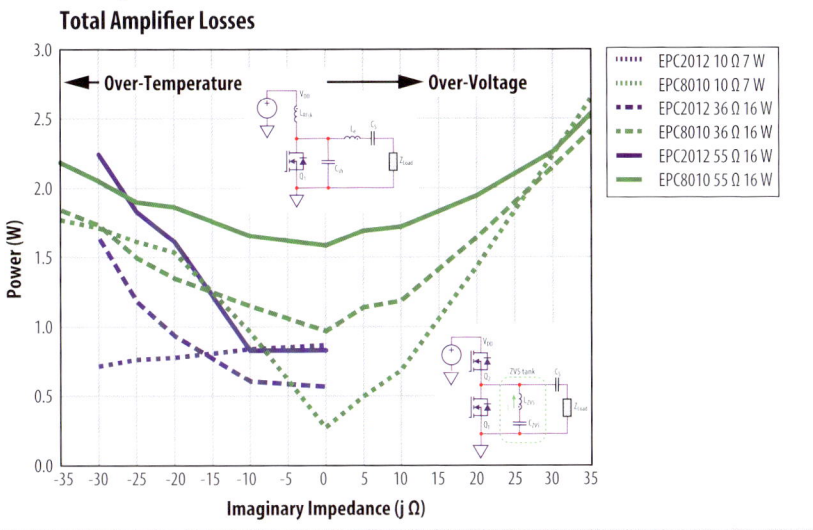

Total Amplifier Losses

Comparison Between Single-Ended ZVS Class D and Class E Amplifiers

The measured amplifier power loss results for both the class E (blue) and ZVS class D (green) amplifiers operating with a varied load impedance range is given in this graph for the eGaN FETs. The effect of imaginary impedance is plotted for various reflected load resistances to highlight the imaginary impedance range capabilities of each of the topologies when complying with the A4WP class 3 standard.

The two higher reflected load resistance ranges were selected based on full power delivered to the load, and the third value at approximately half of full load power. It is notable that the ZVS class D amplifier exhibits a relatively flat efficiency response over the entire imaginary impedance range compared to the class E. This is due to the low output impedance of the ZVS class D amplifier that improves robustness to changes in the coil impedance.

Differential-Mode ZVS Class D Amplifier Class 3 Experimental Results

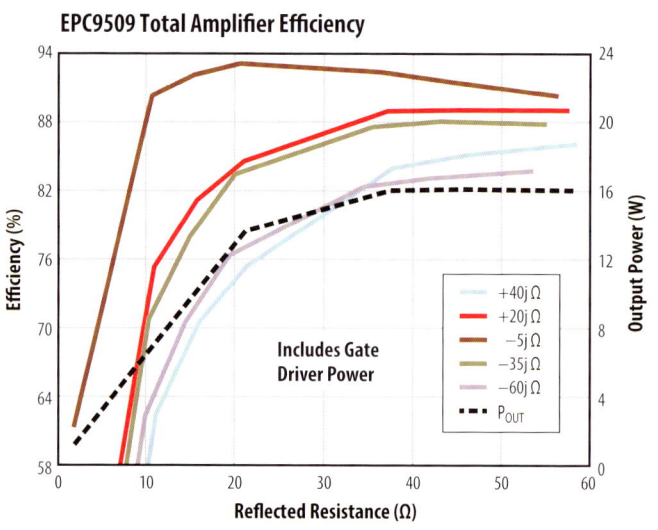

EPC9509 Total Amplifier Efficiency

Differential-Mode ZVS Class D Amplifier Class 3 Experimental Results

The measured amplifier efficiency for the differential-mode ZVS class D amplifier (EPC9509) is shown operating under a resistance load variation of 1.7 Ω through 57 Ω and for various imaginary impedance load conditions to the A4WP class 3 standard. The efficiency is presented in this case over power loss as the performance is not being compared to that of a MOSFET equivalent, but once the system is tested the performance of the amplifier within the system can be evaluated.

The results show that the amplifier has the ability to operate to the A4WP class 3 standard over an incredible absolute range of 100j Ω without the need for adaptive matching. For this amplifier, the coil-set would only require a one-time re-tuning to deliver power over the entire A4WP-compliant range of 160j Ω. The results also show the effect of the current sensor that shifted the operating reactive impedance down from an expected +50j Ω to +40j Ω and from -50j Ω down to -60j Ω.

Summary of A4WP Class 3 Compliance for Amplifiers

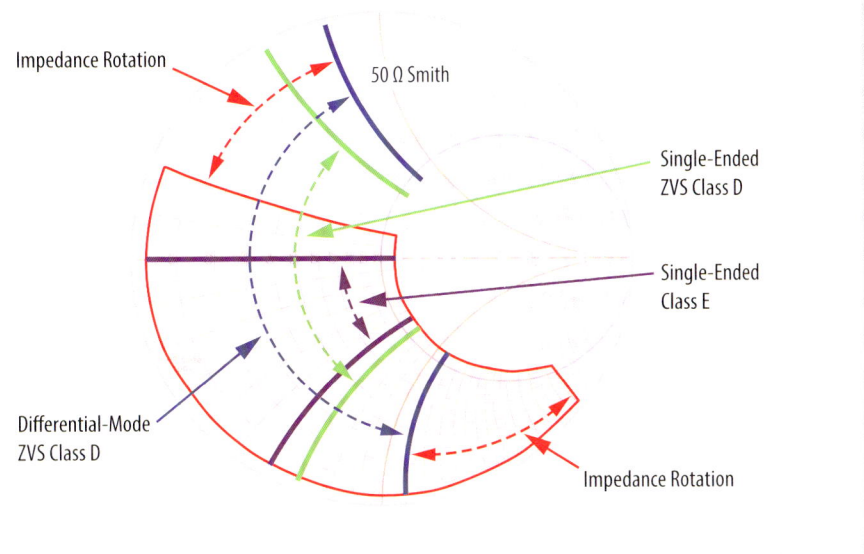

Summary of A4WP Class 3 Compliance for Amplifiers

This chart summarizes the capability of each of the amplifiers tested to the A4WP class 3 standard superimposed on the impedance range of the standard. The purple dashed arc arrow and constant impedance boundaries shows the class E amplifier, the green dashed arc arrow and constant impedance boundaries shows the single-ended ZVS class D amplifier, and the blue-dashed arc arrow and constant impedance boundaries shows the differential-mode ZVS class D amplifier. It is clear from the experimental results that neither of the amplifiers is capable of operating over the entire impedance range.

Retuning the coil is therefore required to rotate the class 3 range to be within the amplifier capability shown by the red arc arrows. Given the range capability of the class E amplifier, the entire class 3 range will need to be broken into at least six discrete steps for full compliance. In the case of the single-ended ZVS class D amplifier only three discrete steps are needed for full compliance. Finally, for the differential-mode ZVS class D amplifier only one discrete step is needed for full compliance.

Some form of adaptive matching circuit will be needed for automatic retuning of the source coil. Discrete adaptive matching circuits offer the simplest solution to retuning the source coil, and each discrete bit will add two tuning set points. This means that a ZVS class D amplifier can potentially yield a lower cost solution through savings in adaptive matching circuitry, since it requires less bits than the class E amplifier.

Single-Ended ZVS Class D Amplifier Class 2 Experimental Results

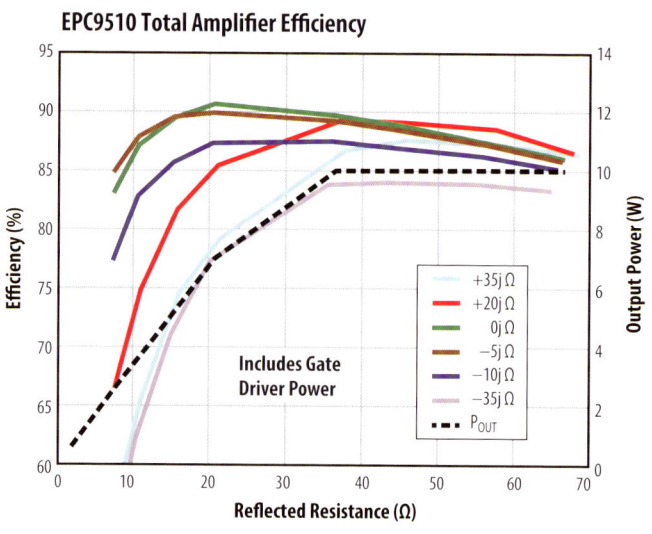

EPC9510 Total Amplifier Efficiency

Single-Ended ZVS Class D Amplifier Class 2 Experimental Results

The measured amplifier efficiency for the single-ended ZVS class D amplifier (EPC9510) is shown operating under a resistance load variation of 6.5 Ω through 65 Ω and for various imaginary impedance load conditions to the A4WP class 2 standard.

The results show that the amplifier has the ability to operate to the A4WP class 2 standard over the entire absolute range of 70j Ω without the need for adaptive matching.

Summary of A4WP Class 2 Compliance for EPC9510

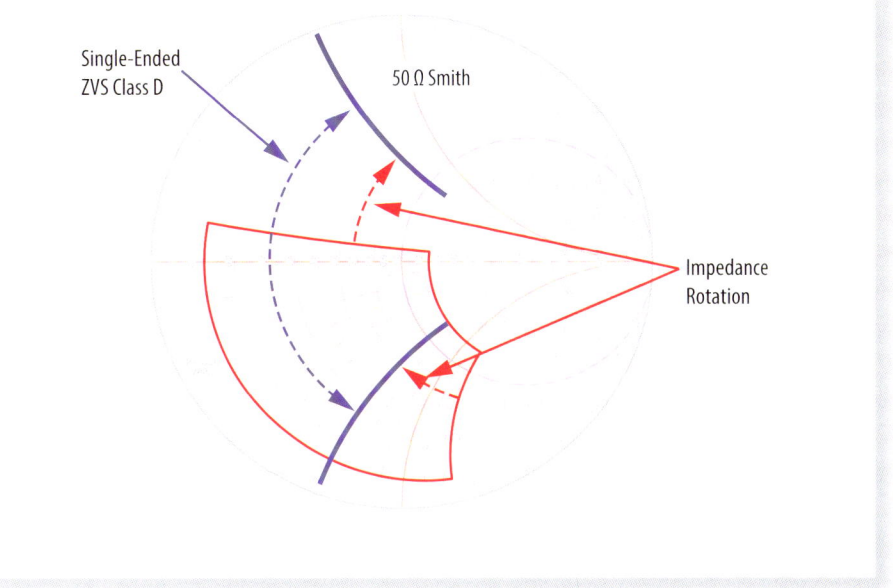

Summary of A4WP Class 2 Compliance for EPC9510

This chart summarizes the capability of the single-ended ZVS class amplifier tested to the A4WP class 2 standard superimposed on the impedance range of the standard. The blue dashed arc arrow and constant impedance boundaries shows the amplifier capability. The red arc arrows show the retuning of the coil for impedance rotation that will bring the class 2 impedance range completely within the capability of the amplifier (hence only arrows pointing in one direction are used as this is a one-time re-tuning of the coil).

System Level Testing

System level performance with actual power transfer will be tested next.

• This will further validate the amplifier performance

• Determines potential issues related to the standard

The wireless power transfer systems to be tested are:

• EPC9114 – Class 2 using the EPC9510 amplifier and one category 3 load

• EPC9113 – Class 3 using the EPC9509 amplifier and two category 3 loads

System Level Testing

Now that the differential-mode and single-ended ZVS class D amplifiers have been tested to the A4WP class 3 and class 2 standards respectively, the amplifiers need to be tested in an actual wireless power system designed to deliver power. This will further validate the performance of the amplifiers and will also be used to check for potential issues.

The wireless power systems to be tested will be the EPC9114 [7.20] using the EPC9510 amplifier to the class 2 standard, and the EPC9113 [7.21] using the EPC9509 amplifier to the class 2 standard. In the case of the class 2 system, only one category 3 device board will be used as load with maximum power delivered of 6.5 W. The class 3 system will use two category 3 device boards that will yield a total delivered power of 13 W.

[7.20] [Online] Available: http://epc-co.com/epc/Products/DemoBoards/EPC9114.aspx

[7.21] [Online] Available: http://epc-co.com/epc/Products/DemoBoards/EPC9113.aspx

Experimental Class 2 A4WP Setup – EPC9114

Category 3 Device

EPC9510 Amplifier

9 mm

Class 2 Source

Experimental Class 2 A4WP Setup – EPC9114

Shown is the experimental setup for the A4WP class 2 system using the EPC9114 [7.22] demonstration kit that comprises an EPC9510 single-ended ZVS class D amplifier, a class 2 source coil and category 3 device board. Nylon standoffs were used to mechanically attach the category 3 device board to the class 2 source coil. The standoffs served to provide a consistent setup ensuring the same location of the device board for each test and the correct distance as prescribed by the A4WP standard between the source and device coils. The setup was designed so that the full A4WP impedance range could be replicated with this coil setup. The amplifier was again operated in the bypass mode in the same manner for the amplifier performance tests.

[7.22] [Online] Available: http://epc-co.com/epc/Products/DemoBoards/EPC9114.aspx

Experimental Class 3 A4WP Setup – EPC9113

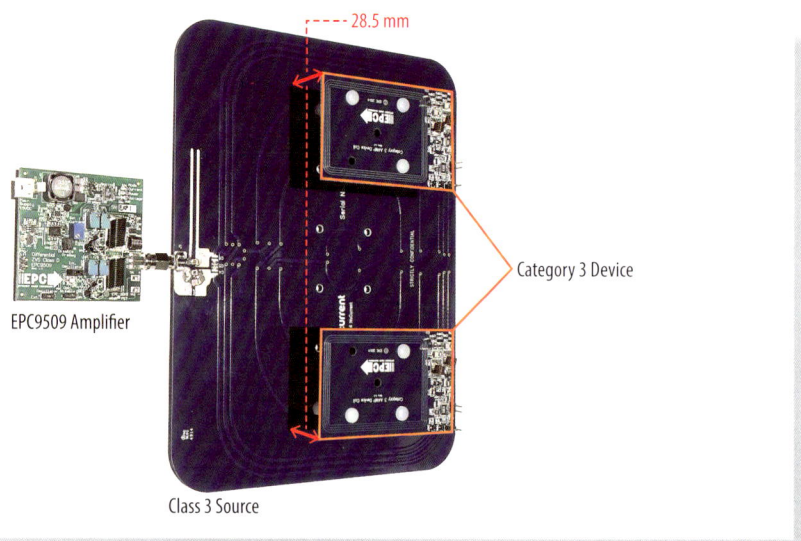

EPC9509 Amplifier

Category 3 Device

Class 3 Source

Experimental Class 3 A4WP Setup – EPC9113

Shown is the experimental setup for the A4WP class 3 system using the EPC9113 [7.23] demonstration kit that comprises an EPC9509 single-ended ZVS class D amplifier, a class 3 source coil, and two category 3 device boards. Nylon standoffs were used to mechanically attach the category 3 device board to the class 3 source coil. The standoffs served to provide a consistent setup ensuring the same location of the device board for each test and to fix the distance between the source and device coils. According to the A4WP class 3 standard, the separation distance between the source coil and device coil is prescribed as 6 mm. It was found that at 6 mm separation between two category 3 devices and the source coil, the source coil became over-coupled and the reflected impedance increased to as high as 200 Ω (when it should only be 56.2 Ω) when each of the device boards were loaded with the minimum resistance (maximum power point). Increasing the distance to 28.5 mm corrected the problem so that the reflected impedance fully complied with the A4WP class 3 standard. This kind of issue and possible corrective actions was presented in chapter 2.

The impedance range that the amplifier was capable of driving was replicated with this coil setup. The amplifier was again operated in the bypass mode in the same manner for the amplifier performance tests.

[7.23] [Online] Available: http://epc-co.com/epc/Products/DemoBoards/EPC9113.aspx

Experimental Class 2 Reflected Impedance Verification

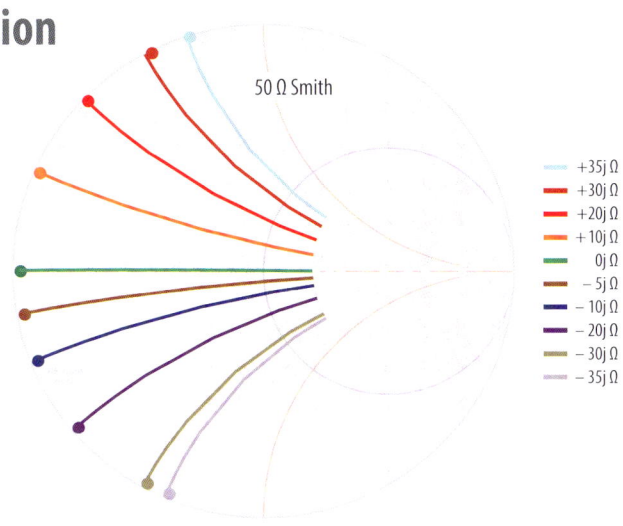

50 Ω Smith

— +35j Ω
— +30j Ω
— +20j Ω
— +10j Ω
— 0j Ω
— − 5j Ω
— − 10j Ω
— − 20j Ω
— − 30j Ω
— − 35j Ω

Experimental Class 2 Reflected Impedance Verification

The coil impedance range under specific operating conditions must be known to be able to verify the performance of the system. The first step was tuning the source coil without the device present. For each change in reflected reactive impedance, the tuning capacitor values on the source coil were changed as indicated by the dots on the Smith Chart, which was easily verifiable using a VNA. Adding the device as an open circuit and decreasing the DC load resistance in discrete steps would primarily increase the real component of the reflected impedance until the maximum was reached near the center of the Smith Chart. The impedance was determined using the analysis methodology described in chapter 2. Additional analysis was used to predict the operating voltage of the amplifier for the various load conditions that were then used to verify the impedance points of the experimental setup.

Experimental Class 3 Reflected Impedance Verification

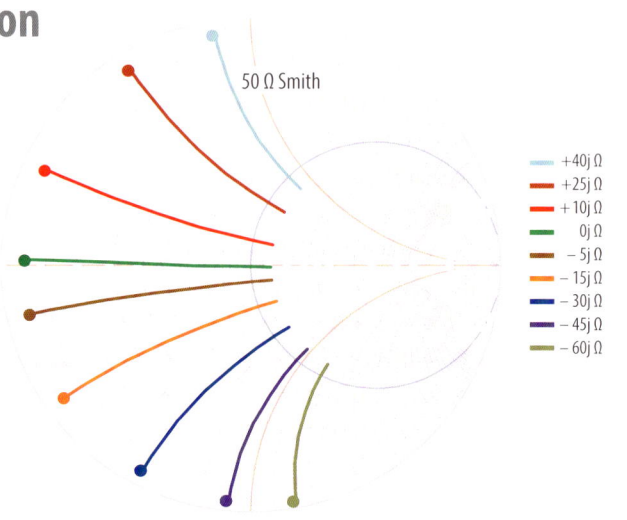

50 Ω Smith

- +40j Ω
- +25j Ω
- +10j Ω
- 0j Ω
- − 5j Ω
- − 15j Ω
- − 30j Ω
- − 45j Ω
- − 60j Ω

Experimental Class 3 Reflected Impedance Verification

The coil impedance range under specific operating conditions must be known to be able to verify the performance of the system. The first step was tuning the source coil without the device present. For each change in reflected reactive impedance, the tuning capacitor values on the source coil were changed as indicated by the dots on the Smith Chart, which was easily verifiable using a VNA. Adding the first device as an open circuit and decreasing the DC load resistance in discrete steps would primarily increase the real component of the reflected impedance to the mid-point. Adding the second device as an open circuit and decreasing the DC load resistance in discrete steps would continue to increase the real component of the reflected impedance until the maximum was reached near the center of the Smith Chart. The impedance was again determined using the analysis methodology described in chapter 2. Additional analysis was used to predict the operating voltage of the amplifier for the various load conditions that were then used to verify the impedance points of the experimental setup.

EPC9114 Class 2 System Efficiency

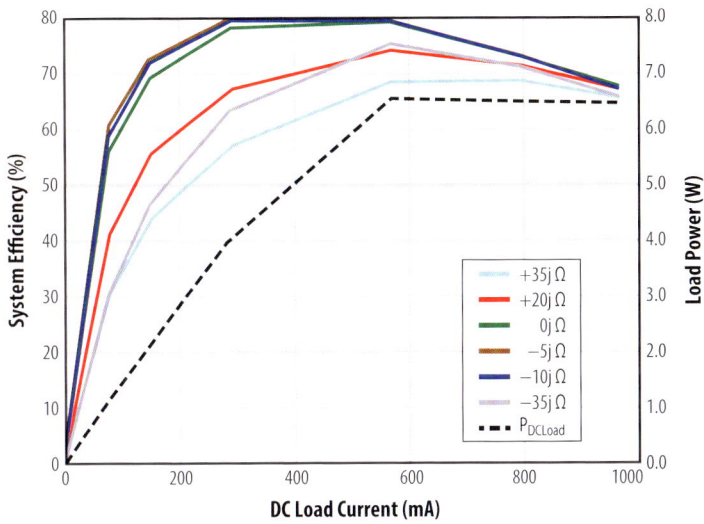

EPC9114 Class 2 System Efficiency

The experimental system efficiency is shown in this graph for the entire class 2 impedance range, and includes gate driver and oscillator power. It is clear that not only does this system fully comply with the A4WP class 2 standard but exhibits excellent efficiency for such a system. This is due to several factors, such as the high efficiency of the amplifier and the correct pairing of the source and device coils for these specific operating conditions. The graph also shows the power delivered into the unregulated DC load by the black dashed trace.

EPC9113 Class 3 System Efficiency

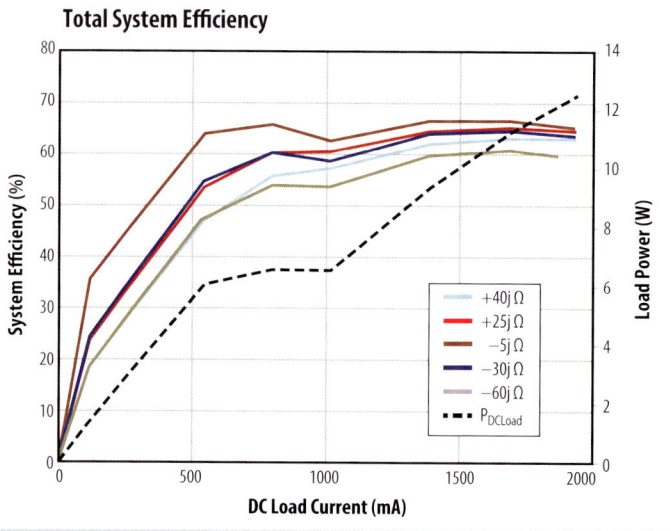

Total System Efficiency

EPC9113 Class 3 System Efficiency

The experimental system efficiency is shown in this graph for the reactive impedance range from -60j Ω through +40j Ω, and includes gate driver and oscillator power. The graph clearly shows when the second device is added with the dip at 1000 mA. This is due to the system being more efficient at delivering 6.5 W with a higher voltage and lower current, than at full current. Adding the second device has a smaller impact to the system, as it is already operating in power limit mode not due to the class 3 requirement, but due to the category 3 device requirements. The results also show that the system is capable of operating over the same impedance range determined for the amplifier using the emulating load. The power delivered into the unregulated DC load is shown by the black dashed trace.

Disappointing is the low system efficiency for this setup, despite using the same category 3 device for this setup as was used in the class 2 setup. The first factor that contributes to low efficiency is the large distance between the source and device coils to achieve the required reflected impedance. The second factor is the high source coil inductance that requires a high reactive energy to maintain the magnetic field. This high reactive energy leads to high losses in the coil. The design of the specific source coil used in this experiment allows the dual-interleave winding configuration to be connected in either parallel or series. The experiment used the series connection, and re-configuring the coil to be connected in parallel may significantly increase the efficiency of the system by addressing the reduction in source coil inductance, and thus reducing the reactive energy required to generate the magnetic field. Additional experimentation is required to verify by how much the efficiency will improve. Connecting the source coil in parallel will unfortunately again cause a deviation from what is prescribed in the class 3 standard.

System Level Observations

Class 2:
- High efficiency
- Full A4WP compliance

Class 3:
- Moderate efficiency
- Not fully A4WP compliant -> requires additional tuning circuitry

Class 3 losses dominated by coil. Coil improvements:
- Lowest possible ESR (at 6.78 MHz)
- Lowest L_{Source}
- Match coupling to device

System Level Observations

A single-ended EPC9510 ZVS class D amplifier based on the 100 V rated EPC2107 was experimentally verified to the A4WP class 2 standard and found that it was fully compliant over the entire impedance range and exhibited high efficiency.

A differential-mode EPC9509 ZVS class D amplifier based on the 60 V rated EPC2108 was experimentally verified to the A4WP class 3 standard and found that it was compliant over a reactive impedance range from -60j Ω through +40j Ω. This means the remaining absolute 60j Ω range would need to be achieved by re-tuning the coil, and will be the subject of the next chapter. The class 3 system tests revealed a low efficiency that is mainly attributable to the incompatibility between the class 3 coil and category 3 device coils as prescribed in the A4WP standard. This can be corrected to yield efficiencies that are comparable or perhaps exceed that of the class 2 system. Corrective actions can include a reduction in the source coil ESR, a reduction in the inductance, and improved pairing of the source and device coils.

CHAPTER 8:

Adaptive Matching - Expanding the Convenience Factor for Wireless Power

The imaginary impedance range is too wide for efficient amplifier operation. The system needs a fast and automatic method to adapt to changing coil parameters, which:

- Improves wireless system efficiency

- Reduces output voltage requirements for the amplifier

- Helps keep EMI generation down

- Addresses convenience factor for wireless power

Required for full A4WP Compliance

Why Adaptive Matching?

Experimental results of testing the class E and ZVS class D amplifiers according to the A4WP class 3 standard has revealed that neither of these amplifiers can operate efficiently and fully comply with the standard. Some form of adaptive matching is needed to retune the source coil to cover the entire impedance range.

Adaptive matching is used to improve coil efficiency as it effectively narrows the impedance range of the coil and brings it back to operate closer to resonance. Also, adaptive matching reduces the output voltage requirement for the amplifier, allowing the resonant circuit to function more optimally at this task.

Retuning the coil also helps keep EMI generation down as the further the system needs to operate from resonance, the lower the quality factor of the tuned coil circuit becomes, and the higher the probability that unwanted frequency sources can impact the coil. This is because the amplifier, which is the main source of EMI, needs to take over the function of the resonant circuit used to increase the coil voltage.

What is Adaptive Matching?

Adaptive matching retunes the source coil's imaginary impedance (results in rotation on Smith Chart)

Various methods can be used to implement adaptive matching, such as switching in various discrete capacitors

What is Adaptive Matching?

In simple terms, adaptive matching is a circuit that monitors the operating conditions at the source coil and operates to change the series capacitance value of the tuning circuit, thereby changing the resonant frequency of the tuned source coil. This effectively rotates the imaginary impedance on the Smith Chart. Various methods can be employed to achieve adaptive matching, the most popular is to switch in and out various discrete capacitors.

Discrete Adaptive Matching Overview

A digital 2-bit adaptive matching circuit results in four settings:

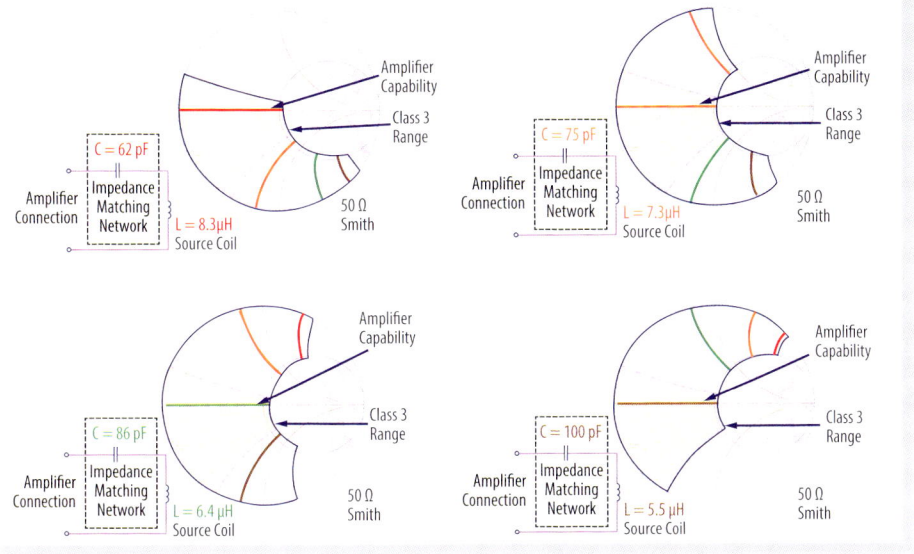

Discrete Adaptive Matching Overview

The adaptive matching process will be examined in detail using a digital two-bit approach that yields four set points. First, we start with an unloaded A4WP class 3 coil and tune it to on-resonance. In this example, the series capacitor required is 62 pF when the coil inductance is 8.3 µH, and is shown by the red trace on the top left graph. Under these conditions an amplifier capable of operating from -30j Ω through +10j Ω can be used to operate in the first quadrant of the class 3 range shown by the yellow area. The orange, green and brown traces fall outside the amplifier operating capability range and represent specific coil impedances for which the coil will need to be re-tuned.

If, for some reason the coil impedance is decreased to 7.3 µH and the tuning capacitor is kept the same, shown by the orange trace on the top left graph, then the amplifier will no longer be capable of operating the load and complying with the standard. Efficiency of the system will drop and the coil must be re-tuned by changing the capacitance to 75 pF (shown in the top right graph orange trace). This re-tuning rotates the impedance, including that of the standard. The amplifier capability remains unchanged, but now it is capable of operating the coil in the second quadrant of the class 3 range.

Discrete Adaptive Matching Overview - *continued*

This process can be repeated where the coil impedance now decreases to 6.4 μH and the tuning capacitor is kept the same (62 pF), shown by the green trace on the top left graph. With the decrease in the coil impedance the amplifier will not be capable of operating the load and comply with the standard even if re-tuned using the 75 pF capacitor. Efficiency of the system will drop further and therefore the coil must be more aggressively re-tuned by changing the capacitance to 86 pF (shown in the bottom left graph green trace). This retuning rotates the impedance, including that of the standard still further. The amplifier capability remains unchanged but now it is capable of operating the coil in the third quadrant of the class 3 range.

Finally, if the coil impedance drops still further to 5.5 μH, and using the original capacitor of 62 pF, shown by the brown trace on the top left graph, then the coil must be re-tuned, yet again, by changing the capacitor to 100 pF (shown in the bottom right graph brown trace). Again, the amplifier capability remains unchanged except now it is capable of operating the coil in the fourth quadrant of the class 3 range.

With all these re-tuning steps, the entire class 3 range can be covered. It should be noted that although a specific inductance value was used in this example, the actual inductance in a practical circuit could be any value within that range, and even outside that range to some extent. The re-tuning process is meant to bring the tuned coil impedance within the range that the amplifier is capable of driving.

Adaptive Matching – Discrete Capacitors

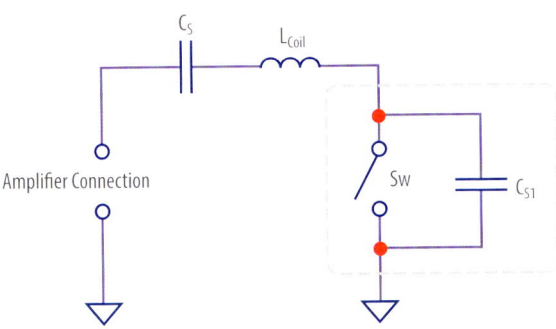

Adaptive Matching - Discrete Capacitors

Now that the adaptive matching process has been explained, we need a method of implementation. In its simplest form, a discrete capacitor can be switched in and out. This yields a one-bit adaptive matching network with two possible states. In the figure shown, an open switch yields lower effective capacitance for the tuning circuit, and a closed switch yields an increased capacitance circuit.

Using the previous example, the capacitance needs to change from 62 pF to 75 pF. With the switch closed, the value of C_s is the only capacitor in the circuit and must be 75 pF. When the switch is open, the total capacitance needs to be 62 pF and, since the both capacitors are connected in series, the value of C_{S1} needs to be 357.7 pF. This assumes that the switch has no capacitance of its own.

Basic Adaptive Matching Network Cell

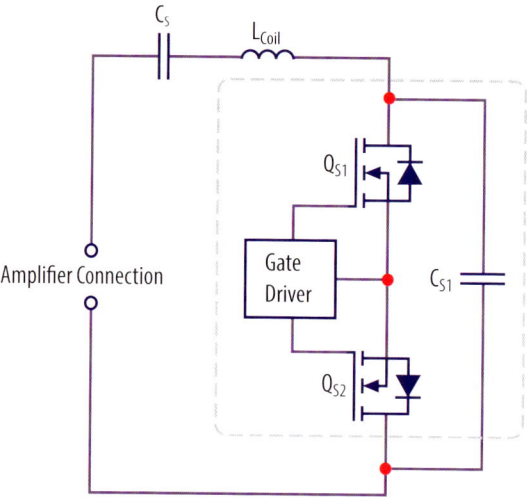

Basic Adaptive Matching Network Cell

Using the basic concept of an adaptive matching circuit cell, a practical implementation can now be derived and a semiconductor switch can replace the ideal switch. The basic requirements for the semiconductor switch are as follows:

- Low C_{OSS} – as this will appear in parallel with the tuning capacitor (C_{S1}).

- Flat C_{OSS} as a function of drain-source voltage (V_{DS}) – any non-linear effects can shift the resonant frequency and mis-tuning the coil.

- High blocking voltage – as the resonant capacitors will increase coil voltage and are exposed to high voltage (which can be as high as 250 V_{RMS}).

- Low $R_{DS(on)}$ – as this will prevent additional losses in the circuit and thus preventing degraded performance.

- dv/dt immunity [8.1]–[8.3] – as the operating frequency and voltage exposure is high.

- High off-state resistance – as this prevents additional losses in the circuit and prevents degraded performance.

Basic Adaptive Matching Network Cell - *continued*

Since the switch needs to block both voltage polarities (the voltage across C_{S1} is sinusoidal) and very few semiconductors can do that, two back-to-back semiconductor switches are used. The most suitable semiconductor switch types for this operation are MOSFETs and eGaN FETs, which meet the criteria for this function. It is further important to note that some amplifier topologies will output a DC offset to the tuned coil circuit, which will need to be added to the blocking voltage capability of the device and resonant capacitors.

Using semiconductor switches necessitates a gate driver. In this function the gate driver has different requirements from those used in the amplifier. This is because it does not switch at high frequency but is required to operate within the high frequency and high voltage environment. The gate driver needs to have the following two characteristics:

- The ability to operate with a high negative voltage (for the negative voltage cycle) with respect to the device source. Alternatively, the gate driver needs to provide isolation.

- High frequency capability and, as such, a low well capacitance to prevent from becoming part of the main resonant circuit.

[8.1] A. Lidow, J. Strydom, M. de Rooij, D. Reusch, *GaN Transistors for Efficient Power Conversion*. Second Edition, Wiley, ISBN 978-1-118-84476-2, chapter 3.4.

[8.2] A. Lidow, J. Strydom, D. Reusch, "GaN – Moving Quickly into Entirely New Markets," *Power Electronics Europe*, Issue 4, June 2014, pp 28–31.

[8.3] T. Wu, "Cdv/dt induced turn-on in synchronous buck regulators," white paper, International Rectifier Corporation.

Multi-Bit Adaptive Matching Network

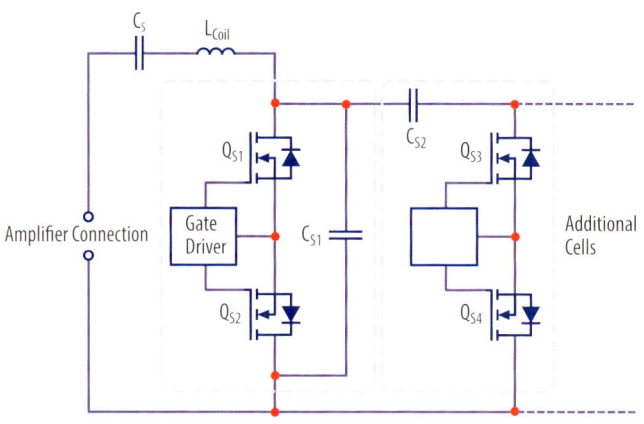

Multi-Bit Adaptive Matching Network

Now that a practical implementation of a single cell adaptive matching circuit has been introduced, the concept needs to be expanded to multiple cells. As was determined from the capability for some of the amplifiers tested, two to three cells are needed to re-tune the source coil to sufficiently cover the entire A4WP class 3 specification range. This requires additional cells that are connected in parallel with the first, as shown in the figure above.

In this case the value of C_S needs to be the maximum capacitance value, and the values of C_{S1}, C_{S2}, through C_{Sn} need to be calculated to yield the values needed for re-tuning. Using the previous example, the value of C_S needs to be 100 pF for use when both Q_{S1} and Q_{S2} are closed. To change the total capacitance to 62 pF with all the cells in the open state, the value of C_{S1} needs to be 163.2 pF. This assumes that the C_{OSS} of all the devices are negligible with respect to the resonant capacitor (C_{Sn}) values. To obtain additional resonant capacitor values, subsequent cell capacitance can be calculated until all the resonant capacitor values can be configured.

Alternative Multi-Bit Adaptive Matching Network

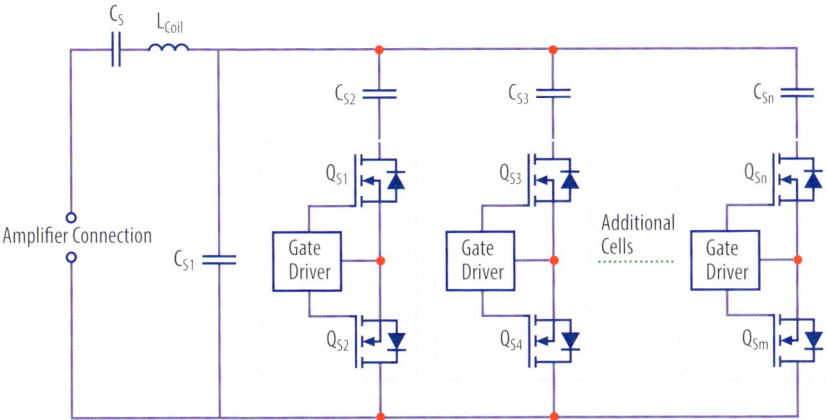

Alternative Multi-Bit Adaptive Matching Network

In the previous implementation of a multi-cell adaptive matching network, three cells are needed to cover the four bits of adaptive matching. These cells are needed because the first cell effectively shunts the capacitor C_{S1}, making it only half as useful. A more effective use of cells is to permanently configure C_{S1} in the circuit and to use two cells to adjust the total resonant capacitance values in a binary manner.

Although this method uses the expensive hardware more effectively, it is much harder to calculate the optimal solution since each bit needs to re-tune the coil by a specific amount. And, this re-tuning is not always going to be a simple linear function. Some compromise can be made if it is determined that the amplifier is capable of operating over a wider imaginary impedance range than the total impedance range divided by the number of re-tuning bits.

Adaptive Matching Challenges

Detection of the amplifiers operating conditions:

- How are coil current, voltage and phase measured?
- Will the detection method introduce losses?
- How does a previous setting affect detection?

Implementation of adaptive matching should consider:

- The impact of parasitics, e.g., C_{OSS}
- A method to eliminate DC offset voltages
- A gate driver with a low well capacitance is needed

Cost for adding adaptive matching network cells:

- Each additional cell multiplies the cost of adaptive matching

Adaptive Matching Challenges

The fundamentals of adaptive matching have been introduced with a basic practical implementation for the power circuit. Many challenges still remain for full implementation because the actual adaptive matching system requires additional circuitry to function, circuitry that includes detection and decision elements.

In a wireless power transfer circuit, operating conditions need to be detected beyond what can be informed by the device. Sensors can affect system operation and EMI generation. In addition, sensors may output information with jitter, which could add further design constraints, and affect the measurement and decision process.

More importantly, will detection methods increase losses in the circuit? Finally, does the decision system need to remember a previous setting after detection of a new state? These questions must be addressed in the design phase to ensure the system can function properly while complying with a standard.

Implementation of adaptive matching circuits introduces parasitic elements such as C_{OSS}, and these must be considered when calculating resonant capacitor values. To make matters more complicated, C_{OSS} is a non-linear function of voltage, and the presence of DC offset voltages can lead to asymmetry and shifts in resonance.

Finally, the gate driver becomes an integral part of the circuit and contributes to parasitic capacitances. Therefore it is important to ensure that the gate driver's well capacitance is low enough not to become part of the power circuit.

The cost of implementing adaptive matching depends on the amplifier's imaginary impedance range capability and the standard it needs to meet. The wider the required impedance range, and the lower the imaginary impedance range of the amplifier, the more cells are needed for adaptive matching, and the higher the system cost.

CHAPTER 9:

EMI for Highly Resonant Wireless Power

EMI Characteristics to be covered:

Related to 6.78 MHz, loosely-coupled A4WP based wireless power systems

- Radiated emissions
- Source side (transmitter) EMI radiation

EMI for Highly Resonant Wireless Power

EMI is a subject that needs to be addressed in any power electronic design, and wireless power transfer is no exception. Wireless power transfer systems are classified as intentional radiators and are therefore burdened with additional regulations [9.1, 9.2] as compared to traditional plug-in power converters. If the spectral content of the energy introduced to the radiator falls outside the transmission bandwidth limits, it will radiate as electromagnetic interference (EMI) [9.3]. This interference is the most difficult to reduce because traditional power circuit EMI abatement techniques will have a significant impact on the performance of the coil and matching circuits. Therefore, in this section, only radiated EMI will be covered for the source (transmitter) circuit.

This transmitter circuit will be examined operating at 6.78 MHz (ISM band), based on the loosely-coupled A4WP, Rezence, standard. Only the engineering aspects will be covered and the process or standards themselves will not. However, knowledge of the relevant EMI standards, which define how EMI is measured, antenna orientation, and antenna distance from source is required. The EMI standards are an integral component of EMI [9.1]–[9.3].

[9.1] FCC Code of Federal Regulations Title 47, Vol. 1, Part 18 B (Industrial, Scientific, and Medical Equipment), 1998.

[9.2] *Electromagnetic Compatibility (EMC)*, European Directive (2004/108/EC).

[9.3] European Norm. EN55011 Group 2 Class B.

Applicable Radiated Emissions Standards

USA:

Radiated emissions limits:

• FCC part 18 consumer (ISM intentional radiator)

Human exposure limits:

• FCC part 1.1310 class B

EU:

Radiated emissions limits:

• EN55011 group 2 class B
• EMC directive (2004/108/EC)

Human exposure limits:

• Recommendation 1999/519/EC
• ICNIRP 2010 (limited adoption)

Applicable Radiated Emissions Standards

Before delving into radiated EMI for wireless power, we first need to understand the limits imposed on the system. The relevant standards are dominated by the US and EU standards.

Three standards govern EMI [9.4]: the first relates to the radiated EMI limits [9.5, 9.6], the second relates to the intentional radiation within the ISM bands [9.6, 9.7], and the third relates to human radio frequency (RF) exposure limits [9.8].

The radiated EMI limits, spanning the frequency range from 6 MHz through 1 GHz, present the greatest challenge to consumer wireless power products falling under class B, which have the lowest limits.

The intentional ISM band radiator standards restrict the frequency bandwidth of the radiated energy but essentially allow unlimited radiated power, with a few exceptions [9.4] in the frequencies targeted for wireless power.

The human RF exposure limits will limit the power radiated by wireless power systems. Not all the standards have been adopted, thus the limits may change in the near future.

Applicable Radiated Emissions Standards - *continued*

Wireless power systems falling under the A4WP standard that operate at 6.78 MHz, an ISM band, in most cases are permitted to radiate unlimited power within the ISM frequency band. Other open ISM band frequencies include, amongst others, 13.56 MHz and 27.12 MHz [9.6]. 13.56 MHz is used for Near Field Communications (NFC) in products such as RFID and card readers that are now included in smart phones. 27.12 MHz is used, under license, for citizen band radio. These other ISM band frequencies happen to be the 2nd and 4th harmonics of 6.78 MHz and some designers have suggested that since those frequencies also fall under ISM band regulations that they do not need to be adequately suppressed in a wireless power application thereby relaxing the design requirements for an EMI filter. While legally this may follow the letter of the regulation, practically it is a path for potential disaster since products such as smart phones already include NCF functionality which may render it inoperable due to the presence of a wireless power feature.

[9.4] J. Roman, R. Paxman, N. Zou, Y. Nakagawa , J. Cho, "Inductive Wireless Power Regulations," *Wireless Power Summit*, Berkeley CA, U.S.A., November 2014.

[9.5] European Norm. EN55011 Group 2 Class B.

[9.6] *FCC Code of Federal Regulations Title 47*, Vol. 1, Part 18 B (Industrial, Scientific, and Medical Equipment), 1998.

[9.7] *Electromagnetic Compatibility (EMC)*, European Directive (2004/108/EC).

[9.8] *FCC Code of Federal Regulations Title 47*, Vol. 1, Part 1.1310 (Radio frequency radiation exposure limits).

Radiated EMI Overview

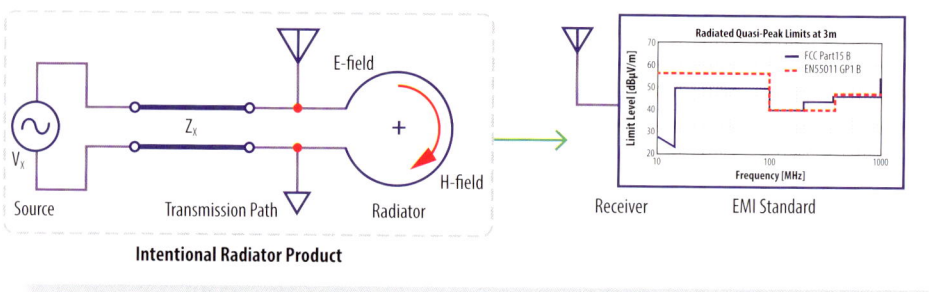

Radiated EMI Overview

A radiated EMI system is comprised of five basic components: 1) a source, 2) a transmission path, 3) a radiator (antenna), 4) a receiver, which is defined as a circuit that can be corrupted, and 5) EMI standards, which set the limits for radiated electro-magnetic energy. Filtering is not considered a component of an EMI system but rather a means to limit the magnitude of EMI radiated within a specific frequency range.

The receiver should not be confused with the intended receiver for EMI testing, but rather any receiver (circuit) that can be corrupted by unwanted frequencies and energy. It is also important to note that a far-field receiver, such as one that would be defined for radiated EMI standards, cannot distinguish whether a source is H-field or E-field generated.

The absence of any one of the EMI system components results in no radiated EMI, however given the presence of the radiated EMI standards and associated receivers, the EMI problem falls solely on the product itself. The further back in the sequence the radiated EMI is addressed in the product, the greater the benefit and easier it becomes to abate the EMI and to comply with the standards which comes with an associated reduction in costs.

The Radiator in Wireless Power Systems

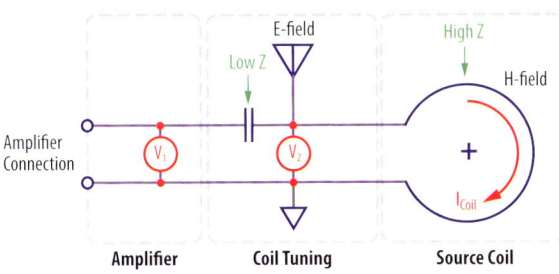

The Radiator in Wireless Power Systems

In a traditional power electronic product, the EMI radiator can be a combination of circuit boards, cable harnesses, and ports. Although this is true for a wireless power product as well, the source coil becomes the main radiator due to its size, construction and function. Hence we will not discuss the EMI radiators from the balance of the wireless system, as it falls into the same category as a traditional power electronic solution, and classic solutions [9.9]–[9.11] can be employed for abatement. It is still an important component of radiated EMI abatement and must be addressed and considered during design and compliance testing as a potential additional source of EMI.

The source coil is simply a large inductor with high impedance. Current is injected into the coil by the amplifier, and assisted by series-tuning the coil with a capacitor to yield low impedance. This series-tuning capacitor is not part of the radiator itself, but is still a crucial element in this part of the circuit. The current in the coil generates an H-field that will radiate; therefore any frequency present in the current will also be present in the H-field. Most of those frequencies will be unwanted and must be prevented from being present in the current prior to entering the source coil. Since there is current present in the source coil inductance, it will have a corresponding voltage (V_2) associated with it, despite having a ground reference. It is furthermore not possible to ground reference the entire coil, and hence the coil will also serve as an E-field radiator.

In addition, voltage (V_1) generated by the amplifier couples directly through the series-tuning capacitor, which provides a low impedance path from the output of the amplifier to the coil and adds to the voltage (V_2) generated by the coil current. This will exacerbate the radiated EMI problem due to the increase in unwanted magnitude and frequency content in the E-field, which is considered the most difficult to abate.

[9.9] On-Semiconductor, "A Solution for Peak EMI Reduction with Spread Spectrum Clock Generators," Appl. Note AND9015, July 2011.

[9.10] Intel, "Design for EMI," Appl. Note AP-589, February 1999.

[9.11] M. J. Schneider, "Design Considerations to Reduce Conducted and Radiated EMI," College of Technology Masters Thesis, Paper 4, 2010, [Online] Available: http://docs.lib.purdue.edu/techmasters/4

EMI Source

EMI is generated by the switching of the power devices

EMI generation between amplifier topologies will be compared:

- Class E
- ZVS class D

Is there a difference between differential-mode and single-ended versions of the amplifiers?

Identify frequencies that pose challenges to mitigate EMI

Analyze potential opportunities to mitigate radiated EMI

EMI Source

The source is the first component in a product's radiated EMI chain, and it is the switching devices in the amplifier that generate the majority of the emissions. Passive components can only act to resonate, filter, and conduct unwanted frequencies. This means that the difference in generated EMI can be largely attributed to the amplifier's topology.

The two topologies that will be investigated, using the same source coil and operating conditions, are class E and ZVS class D. Both topologies have one coil connection that is ground referenced. A popular technique to increase amplifier output power is to use a differential version of the amplifier, so we will explore the question of whether the differential versions radiate lower EMI relative to the single-ended versions. This is important, as both the low- and high-power versions are subject to the same radiated EMI standard limits, making it harder to comply with the standard at higher power levels. The exercise will also be used to identify frequencies within the radiated EMI spectrum that pose significant challenges for compliance.

Lastly, opportunities for mitigating radiated EMI will be investigated. Specifically, we will look at how transmission lines, inherent in a design, contribute to radiated EMI propagation into the source coil.

Single-Ended Class E EMI

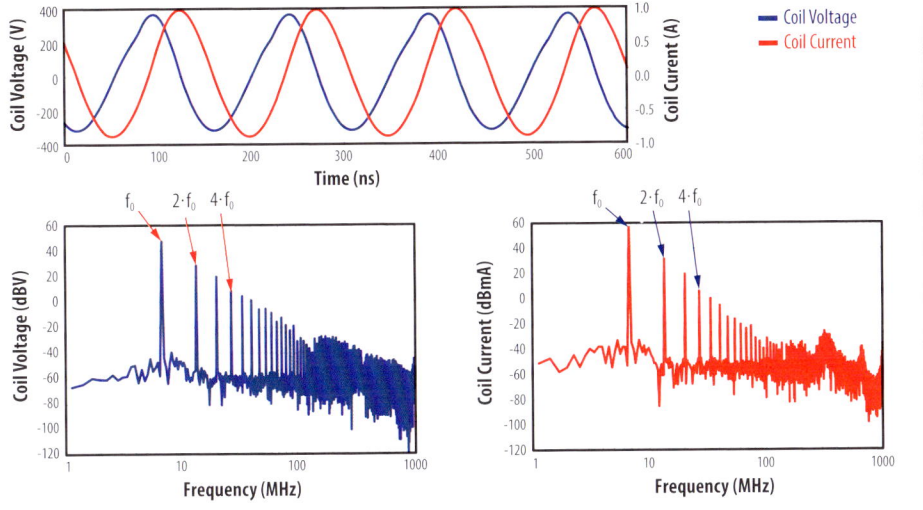

Single-Ended Class E EMI

First we analyze, by simulation, the single-ended class E amplifier driving an equivalent circuit A4WP class 3 compliant source coil set to deliver 14 W into the load. This load power was chosen based on a high-probability operating condition for such a source coil, and the full circuit was simulated in LTspice [9.12].

Special precautions were taken in the simulation to ensure exact timing, thereby avoiding post-simulation Fourier analysis aliasing and transient-related issues. The current in the coil and voltage across the coil, which included the reflected load resistance, were measured, and their frequency content processed.

The time domain results for both the source coil current and voltage are shown in the upper graph, and both appear to be very clean waveforms dominated by the fundamental frequency. By performing a Fourier analysis on the waveforms from 1 MHz through 1 GHz, it becomes clear that both waveforms have significant frequency content.

As predicted, the current harmonic content will be fully present on the voltage waveform, with additional frequencies and corresponding magnitudes. High-magnitudes of even-order harmonics on both the current and voltage waveforms are present. These harmonics are considered the most difficult to reduce and bring within acceptable limits. The difficulty is that even-order harmonics impact the fundamental frequency asymmetrically and can cause power fluctuations. Even-order harmonics are present in a class E amplifier due to the dual resonant frequency structure of the topology.

[9.12] Linear Technology, LTspice Design Simulation and Device Models, [Online] Available: www.linear.com/ltspice

Differential-Mode Class E EMI

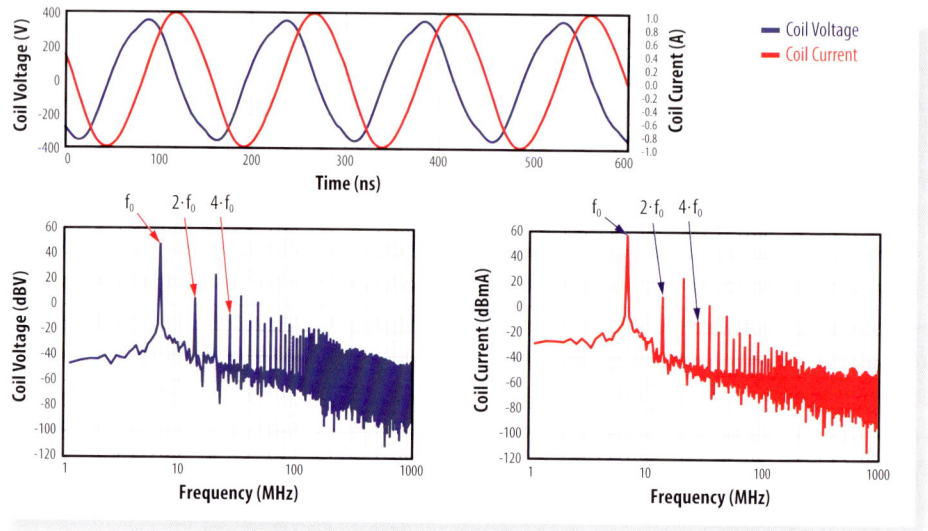

Differential Class E EMI

Next we analyze the differential-mode version of the class E amplifier in a similar manner as with the single-ended class E case. Again the LTspice [9.13] simulation and analysis is performed, where the amplifier drives an equivalent circuit A4WP class 3 compliant source coil set to deliver 14 W into the load. The same special precautions were taken in the simulation to ensure exact timing, thereby avoiding post-simulation Fourier analysis aliasing and transient-related issues.

The time domain results for both the source coil current and voltage are shown in the upper graph and both also appear to be very clean, as in the case of the single-ended amplifier. By performing a Fourier analysis on the waveforms from 1 MHz through 1 GHz, it becomes clear that both waveforms also have significant frequency content.

As predicted, the current harmonic content will be fully present on the voltage waveform with additional frequencies and corresponding magnitudes. Present are high magnitudes of even-order harmonics in both current and voltage. However, in the case of the differential-mode version, the magnitude of the even-order harmonics is lower as compared with the single-ended case. This is due to the differential structure of the amplifier which brings some measure of symmetry into the system. But, as can be seen from the results, it is still insufficient to completely eliminate and ensure radiated EMI compliance. Due to the higher power capability of the differential-mode amplifier, the second-order harmonic issue will be approximately of equal magnitude to that of a single-ended amplifier operating at half the output power as compared with that of the differential-mode amplifier.

[9.13] Linear Technology, LTspice Design Simulation and Device Models, [Online] Available: www.linear.com/ltspice

Single-Ended ZVS Class D

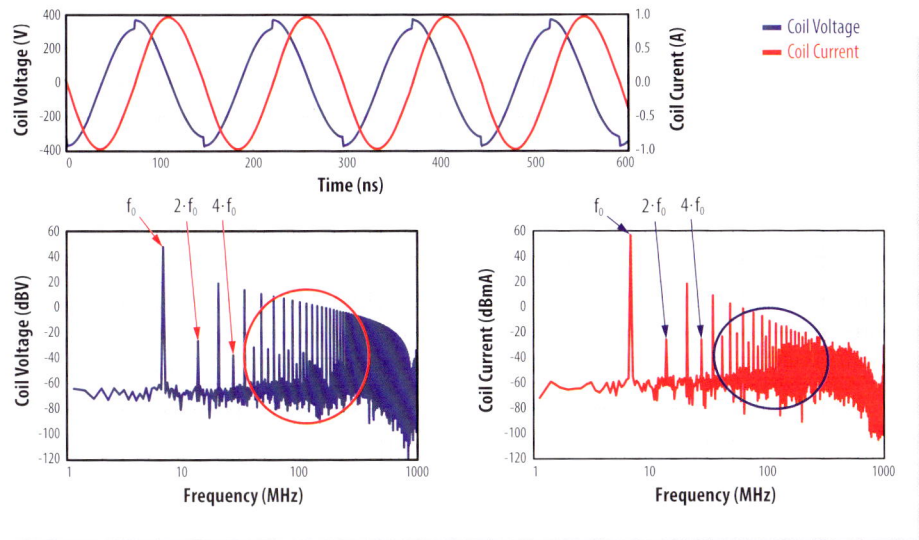

Single-Ended ZVS Class D

Next we analyze the single-ended ZVS class D amplifier. Again, the LTspice simulation and analysis is performed where the amplifier drives an equivalent circuit A4WP class 3 compliant source coil set to deliver 14 W into the load.

The time domain results for both the source coil current and voltage are shown in the upper graph. The current and voltage waveforms appear to be very clean with the exception of the voltage waveform during the amplifier output transition. This is due to the low impedance of the tuning capacitor and amplifier output together with the high impedance of the source coil that allows the output voltage transition of the amplifier to appear on the source coil voltage waveform. Performing a Fourier analysis on the waveforms from 1 MHz through 1 GHz shows that both waveforms also have significant frequency content.

Again, as predicted, the current harmonic content will be fully present on the voltage waveform with additional frequencies and corresponding magnitudes. In the case of the ZVS class D amplifier, the significantly lower magnitude of even-order harmonics in both current and voltage compared to the class E amplifier is notable. This is due to the symmetrical nature of the amplifier and was further found to be independent of variations in duty cycle that may arise from propagation mismatch due to various components in the circuit.

The radiated EMI content in the frequency range from 50 MHz through 300 MHz is higher than the class E amplifier due to the rapid voltage transition of the amplifier output. Since this frequency range is significantly higher than the operating frequency, it should not pose too much of a challenge to reduce for radiated EMI compliance. However, caution should be exercised with EMI mitigation techniques as they can affect the tuning of the source coil.

Differential-Mode ZVS Class D EMI

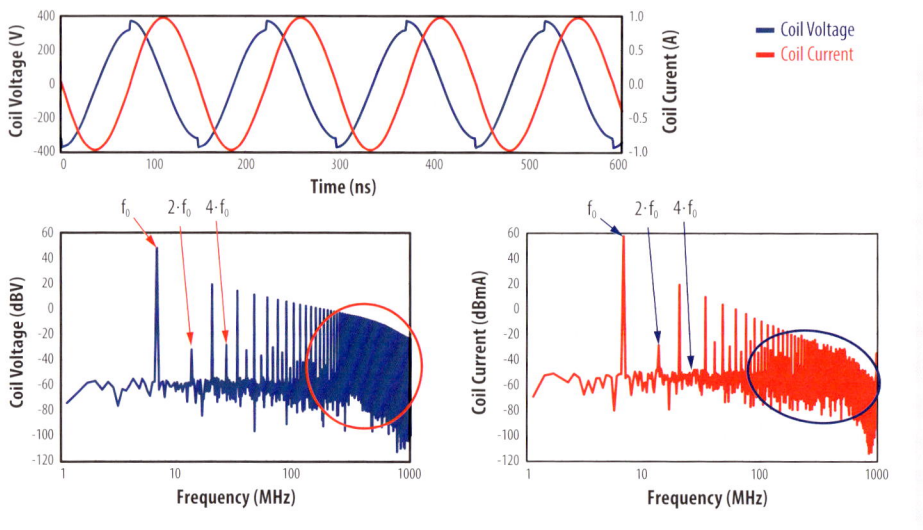

Differential-Mode ZVS Class D EMI

The differential-mode ZVS class D amplifier is the last to be analyzed. Again, the LTspice simulation and analysis is performed where the amplifier drives an equivalent circuit A4WP class 3 compliant source coil set to deliver 14 W into the load.

The time domain results for both the source coil current and voltage are shown in the upper graph. The current and voltage waveforms appear to be the same as those of the single-ended case. Performing a Fourier analysis on the waveforms from 1 MHz through 1 GHz, it becomes again clear that both waveforms also have significant frequency content.

As predicted, the current harmonic content will be fully present on the voltage waveform with additional frequencies and corresponding magnitudes. Similar to the case of the single-ended ZVS class D amplifier, the magnitude of even-order harmonics in both current and voltage waveforms is significantly lower than compared to the class E amplifier. The differential-mode version of the ZVS class D amplifier is also symmetrical in nature, with the same advantages as the single-ended version, such as independence from variations in duty cycle that may arise from propagation mismatch of various components in the circuit. In this analysis, different dead-times and device characteristics were intentionally added to increase the likelihood of even-order harmonic generation, which was not the case for the differential-mode class E analysis.

The radiated EMI content in the frequency range above 300 MHz is lower than the single-ended ZVS class D amplifier due to half the magnitude of the voltage transition at the amplifier output with respect to ground for the same load power. Since this frequency range is significantly higher than the operating frequency, it should not pose too much of a challenge to reduce for radiated EMI compliance. However, caution should be exercised with EMI mitigation techniques as they can affect the tuning of the source coil.

EMI Filter Design Criteria

Traditional EMI filter design techniques will impact coil impedance and can significantly de-tune it.

Therefore an ideal Wireless Power EMI Filter:
- Has very low insertion loss at 6.78 MHz
- Has no gain at 6.78 MHz
- Has high insertion loss outside of 6.78 MHz
- Does not impact the tuned coil impedance (magnitude and phase) at 6.78 MHz

The ideal filter impedance at 6.78 MHz then translates into:

$$Z_{11} = \infty + \infty j \ \Omega \text{ with } Z_{load} = \infty + \infty j \ \Omega \qquad Z_{22} = \infty + \infty j \ \Omega \text{ with } Z_{source} = \infty + \infty j \ \Omega$$

Which is a pass-through of the load impedance at 6.78 MHz

EMI Filter Design Criteria

Adding an EMI filter is inevitable to ensure product radiated EMI compliance, so let's look at the special design requirements for such a filter and where it should located within the configuration of the system. The EMI filter will need to be placed between the amplifier and the tuning circuit for the coil, if no adaptive matching circuit is needed. At this location, traditional EMI filter design techniques will significantly impact the tuned coil impedance as the filter circuit, made up of passive elements, essentially becomes part of the tuning circuit. Therefore an ideal EMI filter design for wireless power needs to have the following characteristics:

- The filter must have no insertion loss at the fundamental frequency, 6.78 MHz in this case. This will keep losses introduced by the filter low. If power loss in the filter is high, then that will impact the entire system efficiency.

- The filter cannot have gain at the fundamental frequency, 6.78 MHz in this case, unless this is intentional as this can severely impact the operating point of the amplifier. Adding gain to the filter also complicates the design and in some cases it may become impossible to find a solution.

- The filter must have high insertion loss outside the fundamental frequency, 6.78 MHz in this case, particularly above the operating frequency. This may be difficult to achieve near the fundamental frequency as practical filter circuits have finite attenuation change rates.

- The filter must not impact the tuned coil impedance so that operation and compatibility with the amplifier remain unaffected at the fundamental frequency, 6.78 MHz in this case.

These design criteria translate into quantifiable terms for the filter at the fundamental frequency of 6.78 MHz. Both the filter input (Z_{11}) and output (Z_{22}) impedances need to be as high as possible; and in the ideal case, will become open circuit or infinity. Practical values of 1k Ω or higher are still acceptable. This essentially makes the filter a load impedance pass through at 6.78 MHz. For all other frequencies, particularly above the operating frequency, the filter impedance change proportionally based on the attenuation characteristics of the filter.

EMI Filter Design Challenges

- **High harmonic content near the fundamental frequency** – increases the order of the filter required to meet the target attenuation

- **Selecting the applicable impedance for the filter** – complicated by the variable impedance of the load and in some cases the amplifier

- **Minimal impact to the tuned coil at 6.78 MHz** including impedance inversion

EMI Filter Design Challenges

The major design challenges for a wireless power EMI filter start with harmonics close to the fundamental frequency. If the magnitudes of the 2nd and 3rd order harmonics are high, then the required filter order will increase to meet the attenuation specifications. The higher the filter order the higher the cost, complexity of design, and difficulty to meet the wireless power special design requirements. From the analysis comparison between the class E and ZVS class D amplifier EMI generation, the class E amplifier generated significant 2nd and 3rd order harmonic content essentially adding one or two filter orders to the EMI filter design compared to that needed for a ZVS class D amplifier.

Most designers require a load (R_L) and source (R_S) impedance to design a filter. For wireless power this is not a simple parameter to specify. The load impedance for a wireless power system varies significantly due to the coil impedance specifications. The choice of amplifier also affects the source impedance. In the case of a class E amplifier, the source impedance is relatively high and was approximately 86 Ω in the example used in chapter 7, and is dependent on the specific design. The source impedance can be reduced by operating the class E amplifier in parallel mode [9.14], but there are limits as to how far it can be reduced. In the ZVS class D case, the source impedance is very low at approximately 1 Ω or less. This makes the ZVS class D amplifier less sensitive to load impedance, until the load impedance also becomes very low, but makes it difficult to choose an applicable filter characteristic impedance. Choosing a low source impedance can lead to strange and unrealizable results. The solution is to always choose the source impedance to be the same as the selected load impedance.

Finally, the EMI filter should have minimal impact to the tuned coil impedance at the fundamental frequency only, 6.78 MHz in this case. In some cases the filter can cause impedance inversion similar to the effect of the "transformer" between the device and source coils. This can cause high power requirements to occur at low impedance, leading to significant increase in losses for the amplifier by increasing supply current requirements.

[9.14] A. Grebennikov, "Load Network Design Techniques for Class E RF and Microwave Amplifiers," High Frequency Electronics, vol. 3, pp. 18-32, July 2004

EMI Filter Design Methodology

- Draft the filter specifications for pass and stop bands
- Determine suitable filter type
- Determine filter impedance specification
- Design the filter
  - Draft a transfer function
  - Convert the transfer function into a passive filter network
- Adapt the filter to the wireless power coil
- Evaluate the filter and revise if necessary

EMI Filter Design Methodology

The procedure to design an EMI filter begins by drafting the filter specifications. The stop-band defines the required attenuation starting at a specific frequency, which most likely will be the 2nd harmonic at 13.56 MHz. The pass-band can either be the minimum specification for the fundamental frequency such as -0.2 dB, which is 6.78 MHz in this case, or the -3 dB point at a frequency slightly above the fundamental such as 7.2 MHz. The choice of pass-band specifications depends on the choice of the type filter.

Next a suitable filter type will need to be selected. There is a strong correlation between the filter type, filter order, and attenuation that can be achieved. Also, some filters may not be suitable for use with wireless power. These aspects will become clearer in the subsequent discussion.

The characteristic impedance (Z_0) for the filter plays an important role in the design, but as mentioned previously, takes on a new meaning for wireless power due to the variable impedance load (Z_{Coil_Tuned}). The choice of filter impedance in this design procedure will focus on using the same value for both the source and load, with the load characteristic driving the value. This will most likely not work for a class E amplifier-based filter design, but for this case it is easier to determine a usable source impedance. The criteria for the selection the filter characteristic impedance will be discussed later.

Once the filter specifications have been drafted the design of the filter can begin. This is a two-step process where first the Laplace transform of the voltage transfer function $H(s)=V_{OUT}(s)/V_{IN}(s)$ is calculated based on the chosen filter. Using the transfer function, it can be converted into an equivalent passive network. This is the most difficult step and can be further complicated by the requirements for wireless power.

Once the filter has been designed, it will need to be adapted for wireless power. If this step is omitted it could result in a filter with a transfer function that meets the specifications but will most likely fail the input (Z_{11}) and output (Z_{22}) impedance requirements. This step is critical to make sure the filter will be capable of fulfilling the radiated EMI reduction requirements while remaining compatible with the wireless power system. In some cases, the filter design may need to be revised and corrected until it meets all the required specifications.

Determination of Filter Specifications

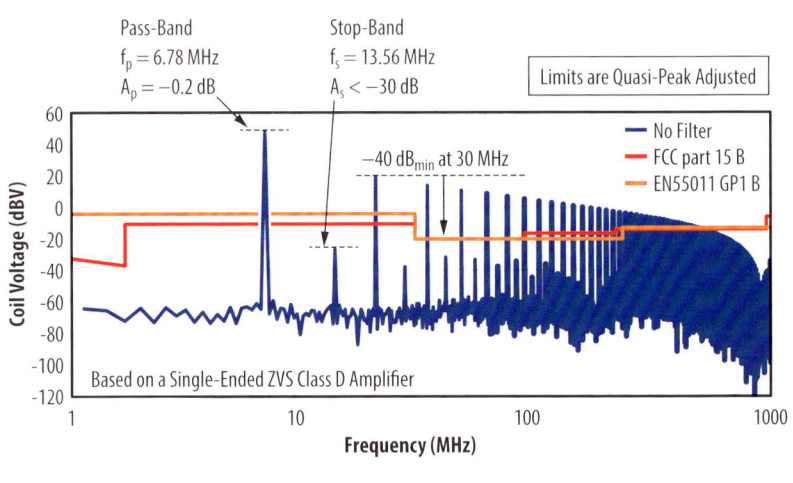

Determination of Filter Specifications

The first step in the design of the EMI filter is to determine the appropriate specifications. This is an important step, as choosing an attenuation that is too high can add complexity to the design and unnecessarily increase the cost of the filter.

The procedure will use a single-ended ZVS class D amplifier, and the reference EMI will be the coil voltage, as it contains higher magnitude and frequency content than the coil current. The frequency content of the coil voltage is shown in the graph (blue trace) together with the adjusted EMI limits (orange and red traces). The adjusted EMI limits were determined from measurement of various similar power level wireless power amplifiers, and the EMI limits adjusted from dBμV to dBV as if it were measured directly across the coil terminals.

In this example 6.78 MHz will serve as a reference for the first frequency specification of the pass-band of the filter, and a maximum attenuation of -0.2 dB is desired. Next the stop-band needs to be specified, and in this example the 2nd order harmonic at 13.56 MHz appears with a magnitude of -30 dBV. Using the standard as a reference point, the minimum required attenuation will be -24 dB. Allowing for margin, the filter will specify -30 dB for the stop-band attenuation.

Review of Filter Transfer Characteristics

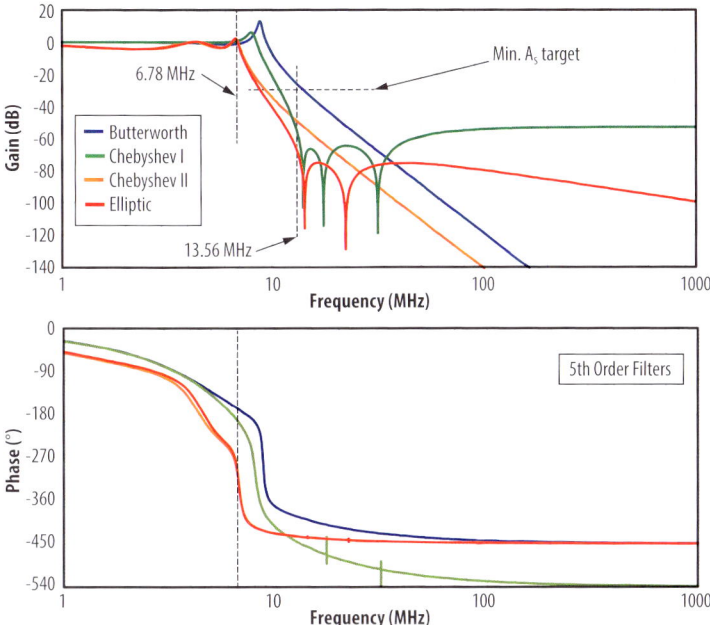

Review of Filter Transfer Characteristics

Using the filter specifications, a low-pass filter type needs to be selected. The transfer functions of four types of filters are shown in the graph [9.15], each as a 5th order realization. The target pass-band and stop-band frequencies have been highlighted using a black dashed line together with the minimum stop-band attenuation of -30 dB. From this graph, it can be seen that the Butterworth filter does not meet specification despite being a favorite due to its flat pass-band characteristic. All the other filters types, Chebyshev Type I and II and Elliptic, meet the minimum stop-band requirement and will further be considered for the design.

[9.15] S. J. Orfanidis, "Lecture Notes on Elliptic Filter Design," Department of Electrical & Computer Engineering Rutgers University, November 20, 2006, [Online] Available: http://www.ece.rutgers.edu/~orfanidi/ece521/notes.pdf

Review of Suitable Filters for Wireless Power

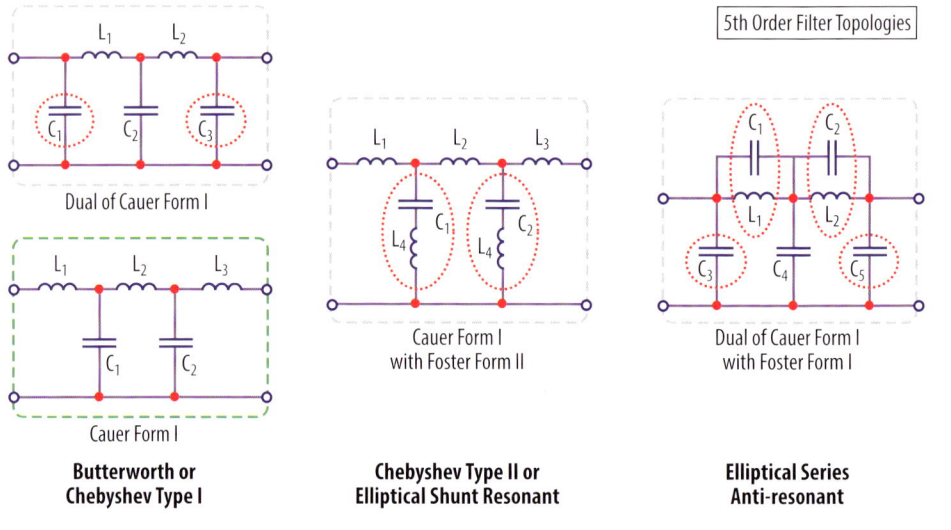

5th Order Filter Topologies

Dual of Cauer Form I

Cauer Form I

**Butterworth or
Chebyshev Type I**

Cauer Form I
with Foster Form II

**Chebyshev Type II or
Elliptical Shunt Resonant**

Dual of Cauer Form I
with Foster Form I

**Elliptical Series
Anti-resonant**

Review of Suitable Filters for Wireless Power

A suitable passive element filter network now needs to be selected from the remaining filter types.

The Butterworth and Chebyshev Type I filters have infinite zeros [9.16] and can be realized using the First Cauer Form I [9.17, 9.18] ladder network or the Dual thereof shown on the left. These filter networks use the least number of components, with the Dual of the First Cauer Form I network having the least number of inductors. The lower inductor count of a filter results in lower insertion losses in the pass-band. Unfortunately, the Dual of the First Cauer Form I network has a capacitor on each end, where one will connect to the output of the amplifier, and the other will connect in parallel with the tuned coil. Since the coil has already been tuned, the impact of the shunt capacitor on the output of the filter will be small, even for capacitor values as high as a few nF's. Unfortunately, the shunt capacitor across the output of the amplifier poses several issues, with the most severe being dramatically increased amplifier losses. So the Dual of First Cauer Form I network is not suitable for wireless power. The First Cauer Form I network, on the other hand, has an inductor on each end. Series inductors are easily tuned out for the coil and thus this implementation of the filter is well suited for wireless power.

The Chebyshev Type II and Elliptic filters have zeros located on the imaginary axis, meaning they have finite zeros. This is used to increase the attenuation from the pass-band to the stop-band at a higher rate than is possible with the Butterworth or Chebyshev Type I filters. This comes at the expense of ripple in the stop band, which in most cases is manageable. The Chebyshev Type II and Elliptic filters can be realized using the First Cauer Form I with Foster Form II modification [9.16], also known as shunt resonant. These shunt elements, however, pose significant design challenges when used in wireless power, as they are not tuned open circuit at the fundamental frequency, 6.78 MHz in this case, and thus interact with the impendence of the tuned coil. Therefore, this implementation is also not suited for wireless power.

Review of Suitable Filters for Wireless Power - *continued*

The Elliptic filter can also be realized using the Dual of the First Cauer Form I with Foster Form II modification [9.19], also known as anti-series resonant. Again, this implementation results in capacitors on each end of the filter in addition to series tuned tank networks. The series tuned networks are also not tuned to the fundamental frequency, 6.78 MHz in this case, but this is less of a problem than the shunt resonant case because each of the series tank networks appears as an inductor at the fundamental frequency. These can be further tuned out with a series capacitor, but now the complexity of the filter is getting too high.

So the only remaining filter network realization suitable enough for the EMI filter is the Chebyshev Type I as a First Cauer Form I realization.

[9.16] J. Grimbleby, "Analog Filter Design," University of Reading, School of Systems Engineering, Electronic Engineering, Lecture SE2A2 Signals and Telecom, [Online] Available: http://www.personal.rdg.ac.uk/~stsgrimb/teaching/filters.pdf

[9.17] J. E. Colgate, "The Control of Dynamically Interacting Systems," PhD Dissertation at the Massachusetts Institute of Technology (MIT), Chapter 6, August 1988, [Online] Available: http://colgate.mech.northwestern.edu/Website_Articles/Dissertation/Chapter_6.pdf

[9.18] A. J. Casson, E. Rodriguez-Villegas, "A Review and Modern Approach to LC Ladder Synthesis," Journal of Low Power Electronics and Applications, Appl. 2011, 1, 20-44.

[9.19] P. Amstutz, "Elliptic Approximation and Elliptic Filter Design on Small Computers," IEEE Transactions on Circuits and Systems CAS-25, No.12 (December, 1978).

Filter Impedance Determination

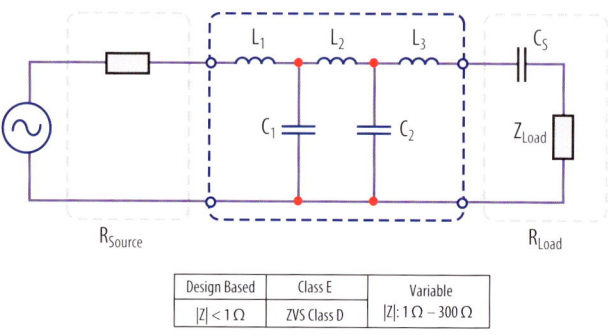

Design Based	Class E	Variable
$|Z| < 1\,\Omega$	ZVS Class D	$|Z|: 1\,\Omega - 300\,\Omega$

Filter Impedance Determination

Before the filter component values can be determined, the characteristic impedance (Z_0) for the filter needs to be specified. For a wireless power system this is not easy to do as the load impedance is highly variable and can have a value anywhere from 1 Ω through 300 Ω, depending on the power level of the system and specific characteristics such as coil size, device to source separation, and number of devices. Also, as previously discussed, the choice of amplifier will have an impact on source impedance. In the case of the class E amplifier the source impedance can be higher or lower than the load impedance. For the ZVS class D amplifier, the source impedance is stable at a low value, typically less than 1 Ω.

The implications for the choice of characteristic impedance of the filter are that a high value will lead to inductors with high values, with high resistance leading to high losses. A characteristic impedance that is too low yields inductors with low values and may cause the filter to draw excessive current from the amplifier when the load impedance is high. So it is important that a value be chosen that yields the lowest losses for the system. Matching the filter impedance to that load or source is not important as the mismatch will be dealt with when the filter is adapted for wireless power in order to meet the input and output impedance requirements.

For the EMI filter design a value of 20 Ω will be used for the characteristic impedance and since this will be a practical implementation, the inductors will need to have a quality factor of 75 or higher.

First Pass Filter Realization

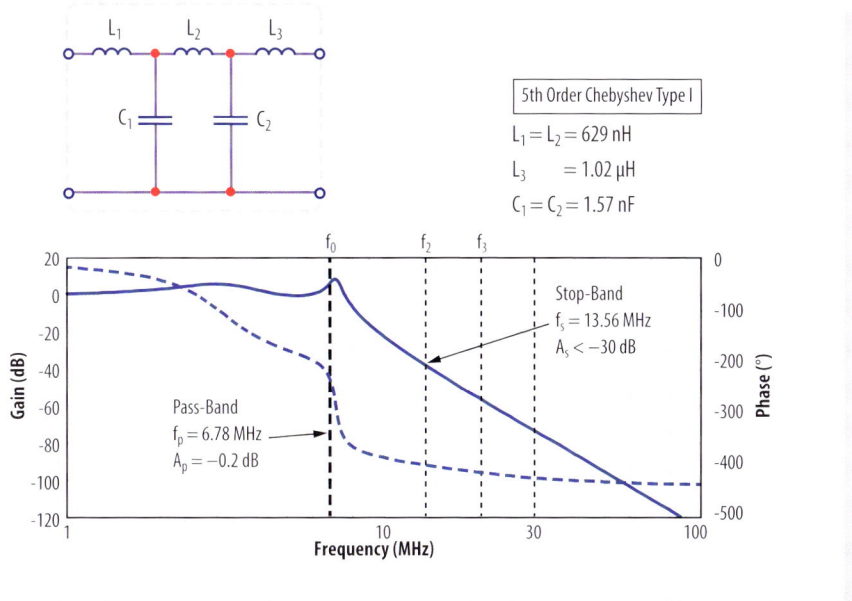

First Pass Filter Realization

The design specifications for the 5th order Chebyshev Type I filter have been determined and can be used to calculate the Laplace voltage transfer function H(s) of the filter as shown in the graph. The transfer function reveals that the filter potentially exceeds the attenuation requirements at the 2nd harmonic.

Using a characteristic impedance of 20 Ω and inductor quality factor of 75, the passive elements can be determined that can be implemented using the First Cauer Form I shown in the upper left. The component values are given in the upper right with values that can easily be selected from standard values and will yield a low loss filter.

Impact of Filter on Coil Impedance

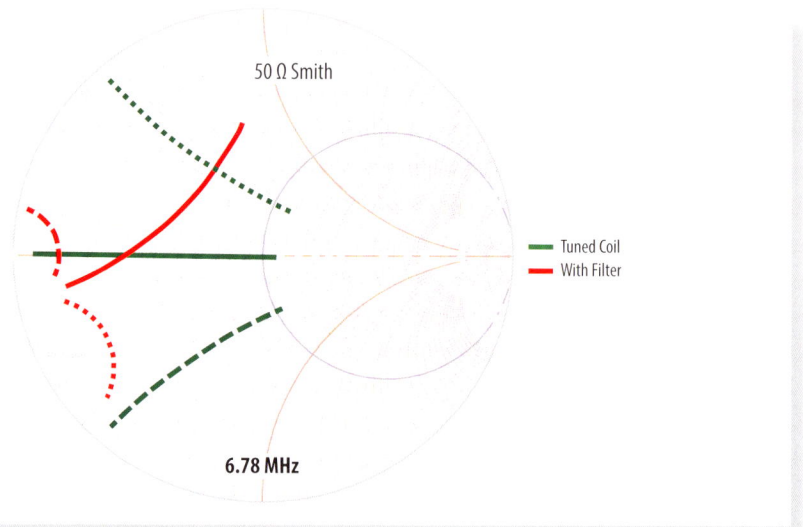

Impact of Filter on Coil Impedance

The designed filter input (Z_{11}) and output (Z_{22}) must now be checked for compatibility with the wireless power coil. The original impedance variation for the coil is shown in green for three reactive impedance settings of -20j Ω, 0j Ω and +20j Ω, each over a real impedance range of 2 Ω through 56 Ω representing a portion of the A4WP class 3 impedance requirements.

The filter is then connected to the coil and the impedance measured looking into the filter towards the coil. The results show a dramatic shift in impedance, shown in red, such that the filter effectively renders the wireless power system useless, and it can no longer deliver power as various power points become extremely difficult to determine and track.

The filter design must therefore be modified to become compatible with the wireless power coil.

Filter Modification for Minimal Impact on Coil

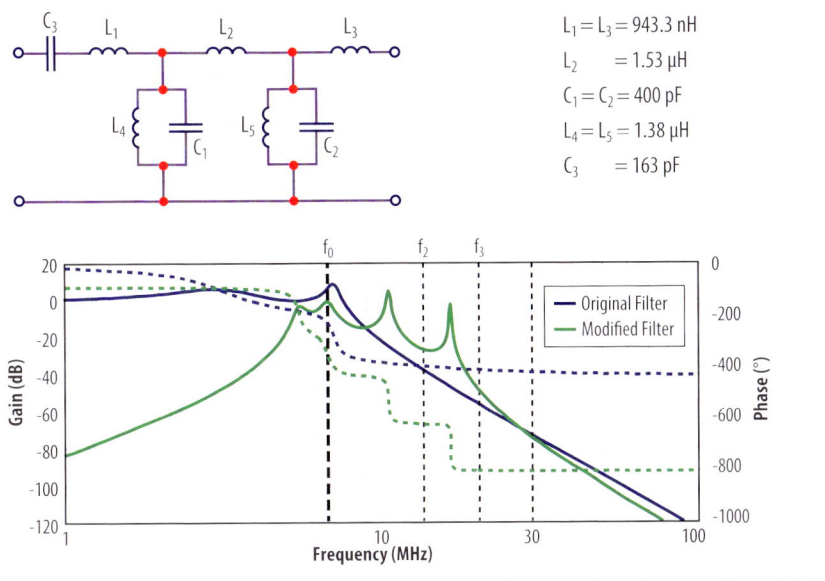

Filter Modification for Minimal Impact on Coil

There are two areas in the filter that require changes in order to re-align the impedance for the wireless power coil. The idea is to make the shunt capacitor branches open-circuit and series inductor branches short-circuit at the fundamental frequency of 6.78 MHz. Making the shunt capacitor branches open-circuit reduces all the series inductor branches to a single inductor and thus only needs to be dealt with once.

Adding an inductor in parallel to the shunt capacitor and tuning it to 6.78 MHz effectively makes that branch an open-circuit at 6.78 MHz only. Doing so will shift the original filter point, and this needs to be re-compensated for by calculating the original branch frequency impedance. So, if that branch was designed for 15 MHz with an impedance of 15 Ω, then then the value of the capacitor needs to be increased so that the impedance at 15 MHz is again 15 Ω, as it originally was without the inductor.

Adding a series capacitor to the series inductor branch, and tuning it to 6.78 MHz, effectively makes that branch a short-circuit at 6.78 MHz only. Unlike the shunt case, adding the series capacitor does not need to be re-compensated as it is effectively connected in series with the series tuning capacitor (C_{series}) of the coil.

Applying this method, the revised schematic of the filter can be derived and is shown in the upper left, with the revised component values in the upper right. The revised filter performance is shown together with the original filter performance. The important markers show that the revised filter still meets the design requirements for the filter. The peaks in the transfer function of the modified filter fall between frequencies generated by the amplifier and should therefore not pose any issues when implemented.

Impact of Modified Filter on Coil Impedance

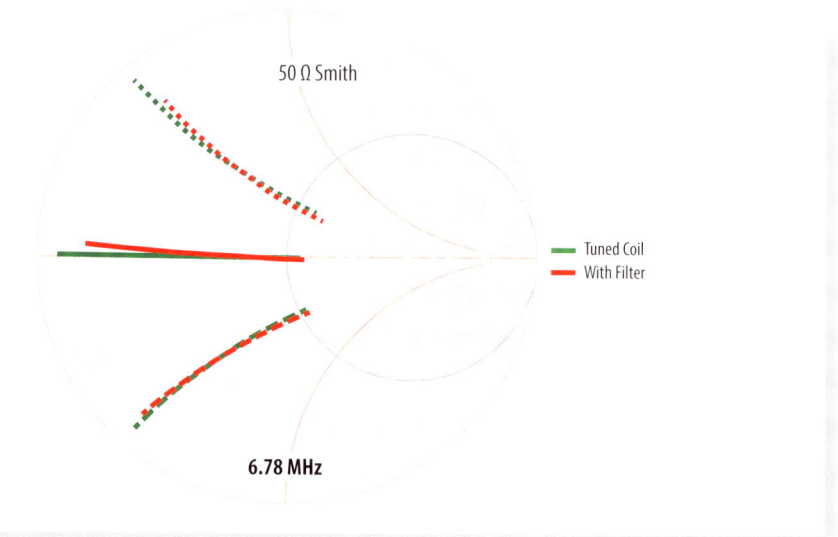

Impact of Modified Filter on Coil Impedance

The modified filter input (Z_{11}) and output (Z_{22}) must now again be checked for compatibility with the wireless power coil. The original impedance variation for the coil is shown in green for three reactive impedance settings of -20j Ω, 0j Ω and +20j Ω, each over a real impedance range of 2 Ω through 56 Ω representing a portion of the A4WP class 3 impedance requirements.

The filter is then connected to the coil and the impedance measured looking into the filter towards the coil, shown in red. The results show that the modified filter has a small impact on the impedance of the coil and mainly increases the real component of the impedance. This is expected as the filter has losses which were included in the modelling.

Modified EMI Filter Performance Review

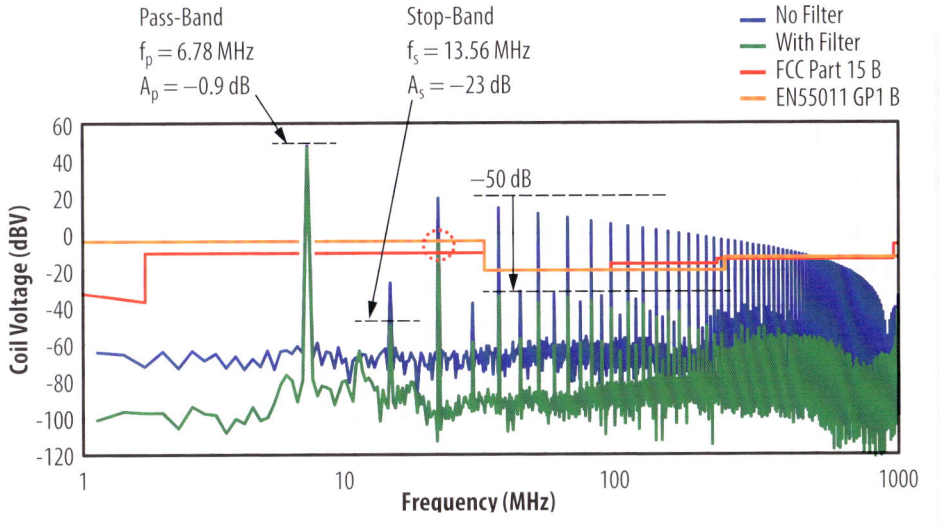

Modified EMI Filter Performance Review

Now that the filter has been designed it needs to be evaluated for EMI performance. Again, the LTspice simulation and analysis is performed where the amplifier drives an equivalent circuit A4WP class 3 compliant source coil set to deliver 14 W into the load.

The graph shows the original single-ended ZVS class D system coil voltage spectrum with no filter attached in blue. The adjusted EMI limits are also shown in the graph with the orange and red traces as presented at the start on the filter design discussion. The coil voltage with the filter attached is shown in green and reveals a significant drop over most of the frequency band.

First, the performance of the filter is checked against the key design parameters. The pass band attenuation is 0.9 dB, which is higher than the 0.2 dB specified. This is due to the quality factor value used for the inductors increasing the impedance and contributing to insertion loss in the pass-band. For the stop-band, an attenuation of 23 dB was achieved, which also fell slightly short of the target of 24 dB minimum, again due to inductor quality factor and the wireless power compatible modifications required.

Relative to the EMI standard, the filter still has good margin of attenuation in the stop-band for all but one frequency at 20 MHz (3rd harmonic). Additional tweaking can correct the filter over the entire frequency band. Overall, most harmonics can be reduced by at least 50 dB. Most importantly, this filter design can operate with the wireless power coil with minimal impact to the impedance.

Addressing Common-Mode Filtering

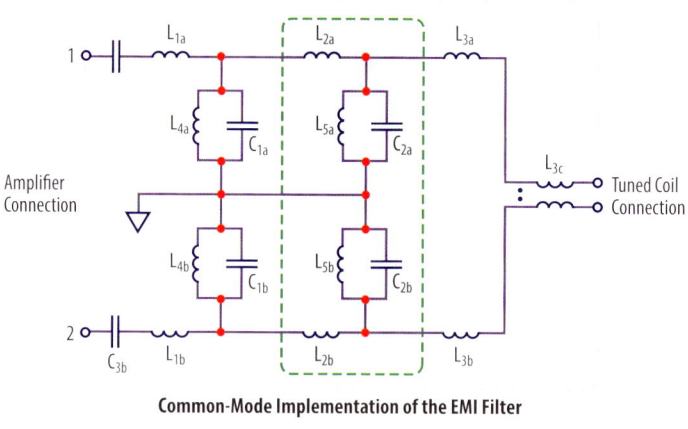

Common-Mode Implementation of the EMI Filter

Addressing Common-Mode Filtering

Up to this point only differential-mode EMI abatement has been discussed. For any electronic circuit, the EMI regulations will need to address common-mode EMI, and wireless power is no exception.

Again traditional common-mode EMI solutions are largely not compatible with the tuned wireless power coils for the same reason as in the differential-mode case. However, the differential-mode filter solution can be adapted to address common-mode filtering. This can be done by creating a second differential-mode EMI filter for the second amplifier connection; essentially a mirror image. Each current path is now filtered so common-mode current can be filtered too. If additional common-mode filtering is required, a common-mode choke (L_{3c}) can be added for the output inductor (L_3). This common-mode choke can add significant inductance to the common-mode path that significantly increases common-mode filtering, and offers some inductance for each of the differential mode paths that can combine with the output inductor (L_3).

The filter order with common-mode capability has now doubled to 10 in this example, and is most likely too high, adding unnecessary cost and complexity to the circuit. The filter order can therefore be reduced to 6 by eliminating one shunt and one series branch from each path shown in the green box. This can be done without loss in the stop-band performance, and pass-band attenuation is expected to be lower. The filter, however, will need to be completely redesigned because the techniques used to derive the passive network differ for a 5th order filter compared to that of a 3rd order filter.

This technique for establishing a common-mode filter works well for differential-mode topology amplifiers, as a third connection becomes available that connects to the ground of the amplifier. For single-ended topology amplifiers the return connection (2) is the ground connection, and the mid-point connection of the filter (formerly the ground connection) becomes floating. This allows the shunt branches to be reduced to two components each.

EMI Filter Component Selection Criteria

Inductors will be the most challenging component to correctly select for the EMI Filter. Key characteristics in the selection process are:

- Self-resonant frequency - This can significantly deteriorate high frequency performance

- Quality factor – Determines the losses and additional impedance shift introduced by the filter

EMI Filter Component Selection Criteria

Practical EMI filter realization will most likely require air core inductors due to their high quality factors relative to equivalent value ferrite core versions. All inductors will exhibit a self-resonant frequency [9.20], so this is an important factor to consider when converting the filter design into a practical implementation. Given that the filter's entire frequency range extends up to 1 GHz, all the inductors used must have a self-resonant frequency that is higher. Typically the lower value the inductor, the higher the self-resonant frequency. If a needed inductor value is not available with sufficiently high self-resonant frequency, then it may become necessary to connect in series two smaller inductors that do have a high enough self-resonant frequency. Self-resonance occurs due to capacitance between the windings that appears in parallel with the inductance. Another technique to increase the self-resonant frequency of the inductors is to increase the separation distance between the windings, yielding a more relaxed coil. This will decrease the inductance, but that can be increased by adding turns.

The quality factor of an inductor is important to consider in the implementation of the filter. This quality factor represents losses in the pass-band, and for filters with zeros, such as the wireless power modified EMI filter, can lead to shifts in impedance, particularly for the shunt branches.

The inductance value tolerance is another important inductor selection metric as it leads to frequency shifting. This is more critical for the shunt branches of the wireless power modified EMI filter, as this shift in frequency will result in impedance shifting that can cause the filter to fail the input (Z_{11}), and output (Z_{22}) impedance requirements.

[9.20] L. Green, "RF-inductor modeling for the 21st century," Electronic Design Magazine (EDN), September 27, 2001, pp. 67 – 74, [Online] Available: http://m.eet.com/media/1142818/19256-159688.pdf

EMI Filter Layout Challenges

Placement of inductors is critical as some may be air core and can easily couple. If air core inductors are used close to each other, then maintain 90° between their field centers to reduce coupling.

High Coupling   Medium Coupling   Low Coupling

EMI Filter Layout Challenges

Of particular importance is the layout of the filter. It is driven by space constraints (as board space comes at a premium) and the unwanted coupling between air-core inductors poses challenges. Inductors placed side-by-side result in high coupling regardless of the relative field direction as shown on the bottom left [9.21].

Inductors placed in-line with each other, shown in the bottom center image, will have lower coupling than a side-by-side placement, since less of the magnetic field can couple. But, this is still not the best practice unless intentional.

The preferred placement for the placement of two different inductors is at 90° with respect to each other, shown in the bottom right image. This minimizes the coupling between them.

In addition, and if possible, the distance between the inductors, shown by the black dashed arrows, should be increased. The larger the distance between the inductors, the lower the coupling.

[9.21] "Square Air Core RF Inductors," AS-Series datasheet, AVX Corporation, [Online] Available: https://www.avx.com/docs/catalogs/assquareaircore.pdf

Transmission Lines in the Design

Transmission lines can be found anywhere in the circuit

These transmission lines can be filters or resonators

Transmission lines affect matching if used to connect the source coil to the circuit, which:

- Will affect tuning
- Can result in unwanted resonances

These transmission lines typically have unknown characteristic impedance

Transmission Lines in the Design

Transmission lines are the second component in the product's radiated EMI chain. Transmission lines can appear anywhere in the circuit by design and, unfortunately, are unavoidable. They are mostly short and of unknown impedance, which can lead them to act as filters or resonators.

Unknown filters and resonators can affect the performance of the amplifier, which is further exacerbated by the wide load operating range required by wireless power systems. This means that these transmission lines can affect source coil tuning if used in the source coil circuit path, and may cause unwanted resonances that can eventually increase radiated EMI.

A Transmission Line Example for Coil Connection

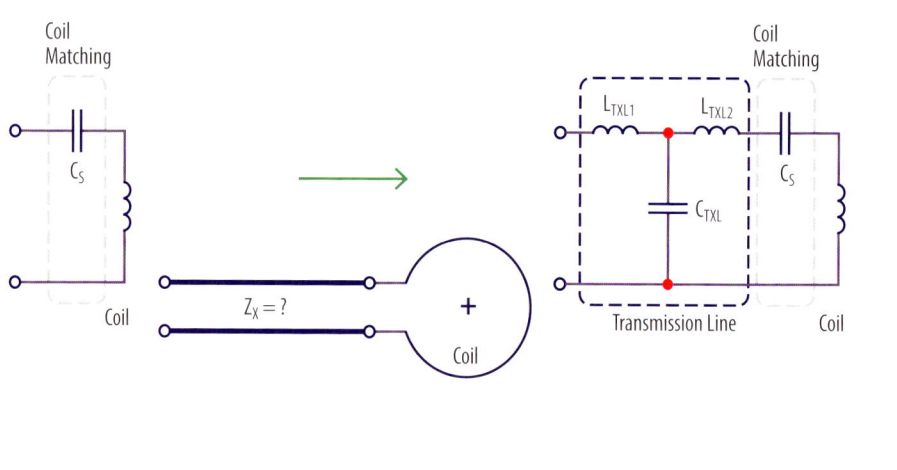

A Transmission Line Example for Coil Connection

An example will be used to understand the impact of a transmission line in a wireless power circuit. In this example, a transmission line has been designed into the circuit to connect the amplifier, located in one corner of a circuit board, to the source coil, with connections in the center of the board near the edge. This is a common design scenario for an A4WP class 3 wireless power system.

The transmission line may, or may not, be designed with a specific characteristic impedance. However, due to the source coil's nominal current requirements, this design approach will lead to printed circuit board traces of significant width that are rated to carry the current and, hence, the characteristic impedance will be 50 Ω or lower.

The source coil impedance has a wide operating range, thus no design is possible to ensure a matched transmission line to the coil over the entire operating reflected load range. The source coil can then be tuned with the inclusion of the transmission line, and one would expect the simplified equivalent circuit shown on the left figure. However, using the simplified equivalent circuit for a transmission line [9.22], the second-order equivalent circuit is actually shown on the right. The introduction of the transmission line capacitance (C_{TXL}) creates a shunt path for current that is essentially tuned to a different frequency as compared to what is desired. This can lead to unwanted resonances, which ultimately will radiate from the source coil.

[9.22] W. H. Hayt, Jr., *Engineering Electromagnetics.* McGraw-Hill, 4th edition, 1981.

Effect of a Transmission Line on Coil Performance

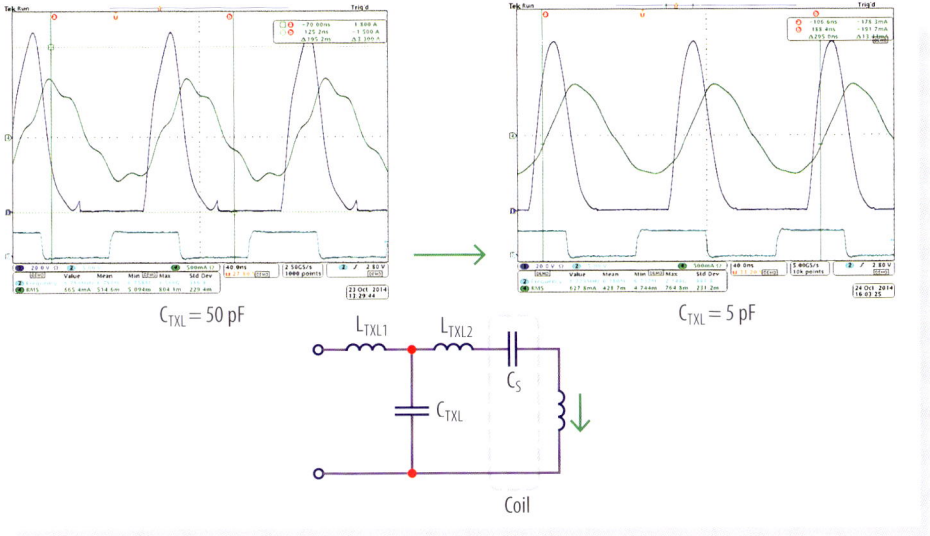

$C_{TXL} = 50$ pF

$C_{TXL} = 5$ pF

Effect of a Transmission Line on Coil Performance

Based on the example presented, an experimental single-ended class E amplifier driving an A4WP class 3 compliant load was tested. In this case, the load included a printed circuit board microstrip transmission line approximately 50 mm in length, and is shorter than what is typically encountered in A4WP class 3 compliant systems. The system was set to operate at the optimal design point that is on-resonance for the cleanest waveforms to deliver 16 W into the load. The measured current in the source coil is shown in the left oscillogram together with the drain voltage of the class E amplifier.

A higher-order harmonic component is clearly visible in the time domain waveform. The transmission line simplified equivalent circuit capacitance was calculated to be approximately 50 pF.

Subsequently, the top conductor of this transmission line was removed and replaced with an insulated wire. In this case, the transmission line simplified equivalent circuit capacitance was reduced to approximately 5 pF. The same circuit was then operated to the same load conditions and re-measured with the results indicating a significant reduction in higher order harmonic content, as shown in the right oscillogram. The importance of understanding the effects of transmission lines in the circuit cannot be overstated as demonstrated in this simple experiment.

Radiated EMI Summary

Analyze the EMI filter requirements to yield the lowest order filter

Filter design requires impedance correction to minimize impact on the tuned coil impedance

Avoid amplifiers that generate high even-order harmonics since they are very difficult to remove

Beware of transmission lines as these can resonate at unwanted frequencies

Radiated EMI Summary

Both class E and ZVS class D amplifier topologies using eGaN FETs were evaluated for radiated EMI generation and propagation. The faster transients of eGaN FETs compared to MOSFETs yield higher frequency generation, particularly when the amplifiers operate with the source coil off-resonance and with non-ideal reflected load resistance. These higher frequencies are beneficial, since the higher the unwanted frequencies are relative to the fundamental, the easier it becomes to filter them out for radiated EMI compliance. The higher frequencies are more important for wireless power systems, since the manner in which the source coil is tuned contributes to creating a low impedance path that propagates energy from the amplifier to the source coil.

Also, it was shown that both implementation modes (single-ended and differential) of the class E amplifier fundamentally generate high magnitudes of even-order harmonics which are very difficult to remove, particularly if the frequencies are close to the fundamental. The ZVS class D amplifier, however, generates significantly lower magnitude even-order harmonics, making it a cleaner choice with respect to radiated EMI.

The design process for an EMI filter was presented and began by defining the requirements for the pass and stop bands. Here the choice of requirements drives the filter order and ultimately the complexity and cost.

The filter design must be adapted for use with wireless power as it is insufficient to only regard traditional transfer function performance. For wireless power EMI filters, it will nearly always be required to adapt the filter to meet the impedance requirements imposed by the coil.

Radiated EMI Summary - *continued*

Amplifiers that generate high magnitudes of low frequency harmonics should be avoided, as the closer the frequency needed to be filtered out is to the fundamental, the harder it becomes to realize the filter. Comparing a class E amplifier to a ZVS class D amplifier, 2nd order harmonic generation is significantly lower for the ZVS class D amplifier, making it a better choice.

The effect of transmission lines in the design of a wireless power product was also investigated in a specific example that showed, due to the mismatch between amplifier output impedance, transmission line characteristic impedance, and source coil reflected impedance, unwanted resonances can be stimulated by the transmission line and radiated by the source coil. The use of transmission lines in the design must therefore be carefully considered and unwanted frequencies must be adequately dealt with prior to being supplied to the transmission line.

CHAPTER 10:

Multi-Mode Wireless Power

The presence of multiple wireless power standards (Qi, PMA and A4WP) results in:

- Confusion for end-users
- The loss of inter-operability

We want universality of wireless power transfer, similar to WiFi
Is inter-operability among the standards possible?

The Current Wireless Power Experience

Wireless power transfer is a relatively new concept that has only recently gained attention as being commercially viable. Unfortunately, there are already a multitude of wireless power standards that have emerged. These competing standards serve to hinder adoption of this technology, as they lead to end-user confusion and loss of inter-operability. What the end-user seeks is something simple, similar to their WiFi experience. No matter where in the world one is, WiFi works. The question now arises that if inter-operability among the various wireless power transfer standards is possible, can that lead to WiFi-like experience for wireless power transfer?

Similarities Between the Wireless Power Standards

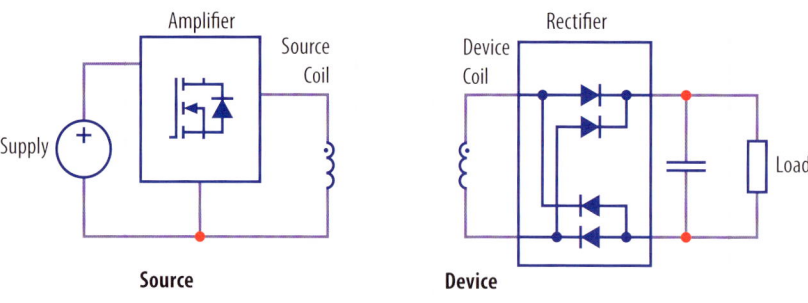

Source   **Device**

Overlap in source power range up to 15 W

Similarities Between the Wireless Power Standards

To begin answering the question of inter-operability, we need to take a look at the wireless power system architecture. This is important to determine similarities between the architectures of systems among the various standards. Also, it is important to know whether there is enough overlap among them to potentially permit them to be used in a multi-mode system.

In general, all the wireless power systems have similar structures made up of an amplifier, a coil-set, and a receiver that includes a rectifier. This general system is shown in the above figure.

In addition, the power of each of these systems needs to be similar. The Qi standard [10.1] can now operate up to 15 W output power and the PMA standard [10.2] can operate up to 5 W output power. The A4WP category 3 device [10.3] is rated at 6.5 W and the category 4 device is rated at 13 W. These similarities between the various wireless standards are enough to set the stage for a multi-mode approach.

[10.1] "System Description Wireless Power Transfer," Vol. I: Low Power. Part 1: Interface Definition, Version 1.2, June 2015.

[10.2] Power Matters Alliance. [Online] Available: www.powermatters.org

[10.3] *A4WP Wireless Power Transfer System Baseline System Specification (BSS)*, A4WP-S-0001 v1.3.1, February 25, 2015.

Differences Among the Wireless Power Standards

• Tightly coupled	• Loosely coupled
• Non-resonant (some)	• Resonant
• Low frequency	• High frequency

We can use the large frequency difference among the standards to further enable a multi-mode approach

Differences Among the Wireless Power Standards

Having determined the similarities among the various wireless power standards, now the differences need to be established. Both the Qi and PMA standards make use of "tight coupling" between the source and device. Tight coupling is required for systems using electromagnetic induction and are therefore primarily non-resonant. However, recently the Qi standard was updated to include some form of resonance to improve operation and efficiency of both the amplifier and system. These systems operate at a relatively low frequency range from 100 kHz through 315 kHz.

The A4WP (Rezence) standard makes use of "loose coupling" between the source and device, and therefore requires resonance to operate efficiently. The A4WP standard operates at a single, narrowband frequency of 6.78 MHz, which is more than an order of magnitude higher than the Qi and PMA standards. This large difference in frequency can be used for automatic coil selection that will further enable an inexpensive multi-mode approach.

Multi-Mode – Coil Selectivity

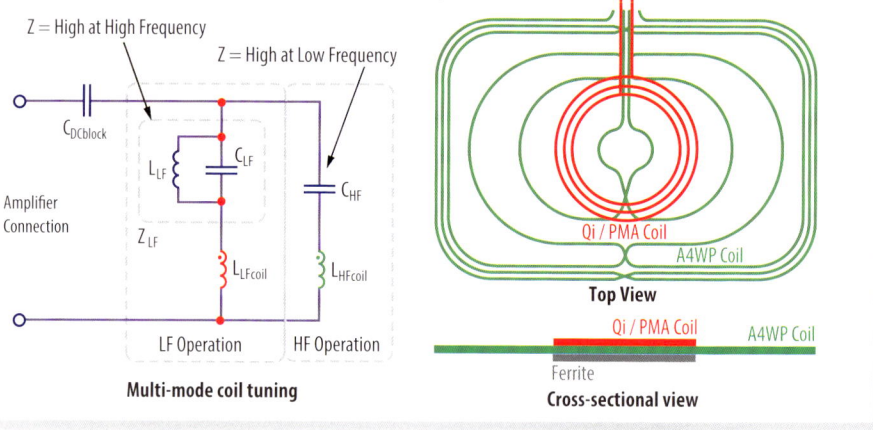

Multi-mode coil tuning

Cross-sectional view

Multi-Mode – Coil Selectivity

Next, we need to take a look at how to implement a multi-mode wireless power system. Proposed solutions begin by having a device that can operate with any of the transmit system formats. In this vision, any multi-mode device could be placed on any standard source and function correctly. Unfortunately, this does not solve the problem of legacy, where there are devices that conform to only one standard. To solve this problem, a multi-mode source is proposed that can accept any device standard [10.4]. This is the approach that is preferred and will be presented here.

The design of a multi-mode source begins with the coils, using one of the standards. In the above figure, an A4WP standard class 3 source coil is integrated with a coil that complies with both the Qi and the PMA standards, as shown on the right. The Qi and PMA coils are very similar, allowing a single coil to be used for either standard. The smaller coil requiring tight coupling is placed on top of the large coil that can operate with loose coupling. Each of the coils can be energized independently, although this is expensive since it requires two amplifiers.

To overcome the need for two amplifiers, the coils can be decoupled and then connected in parallel, as shown in the figure on the left. A parallel resonant tank circuit (Z_{LF}), comprised of a capacitor (C_{LF}) and inductor (L_{LF}), is selected to be resonant at high frequency, 6.78 MHz in this case. This parallel tank circuit becomes high impedance at the high frequency, thus allowing the current to flow only in the high frequency circuit. The parallel tank circuit also prevents the high frequency voltage developed across the low frequency coil (L_{LFcoil}), which is magnetically coupled to the high frequency coil (L_{HFcoil}), from generating a current in the low frequency branch.

For low frequency operation, the capacitor (C_{HF}) will have a high impedance and thus act in the same manner as the resonant tank circuit (Z_{LF}), blocking low frequency current from flowing in the high frequency branch. The inductor (L_{LF}) has a relatively small value compared to the coil inductance (L_{LFcoil}) thereby contributing little to degradation in performance for low frequency operation.

Multi-Mode – Coil Selectivity - *continued*

Now the parallel combination of the two branch circuits can be driven using a single source where only the difference in operating frequency determines which coil will carry current. Finally, since some amplifiers generate a DC offset on the output, a DC blocking capacitor ($C_{DCblock}$) is added. This is only needed if the low frequency coil will not be tuned as per the Qi standard.

One point of note for a multi-mode system concerns a multi-mode source that is paired with a multi-mode device. In this case, the system should default to the A4WP standard so that it allows multiple devices to be powered from the multi-mode source.

[10.4] V. Muratov, "Multi-Mode Wireless Power Systems can be a Bridge to the Promised Land of Universal Contactless Charging," *Wireless Power Summit*, Berkeley CA, U.S.A., November 2014.

Multi-Mode Amplifier

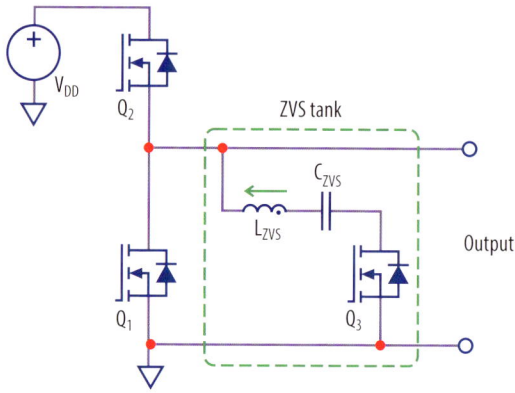

Multi-Mode Amplifier

Now that the coils of the various standards have been integrated, a single amplifier that can drive them needs to be found. Only amplifiers that do not rely on resonance for high efficiency operation can be used. This requirement immediately eliminates the class E amplifier as it will be difficult and expensive to implement a dual resonant frequency mode, and the additional cost is about the same as adding a second amplifier to drive the second coil.

The solution is to use the ZVS class D amplifier, as it is does not rely on-resonance for high efficiency operation, and it has a low impedance output at both frequencies. The only modification to the ZVS class D amplifier required is to switch in and out the ZVS tank circuit used for the zero voltage switching of the switch-node. This must be done since the ZVS inductance used for high frequency will lead to excessive currents at low frequency. Fortunately, this adaptation adds minimal cost to the overall system. Since the ZVS tank disconnect switch (Q_3) does not switch at high frequency and can be driven directly from logic. The EPC2036 [10.5] or EPC2037 [10.6] are well suited for this function as they have a very small footprint of 0.9 mm x 0.9 mm and are very low cost.

eGaN FETs have a proven track record of yielding high efficiency [10.7], [10.8] capability for both operating frequencies (300 kHz [10.7] and 6.78 MHz [10.8]) despite operating in the hard-switching mode at the lower frequency.

[10.5] Efficient Power Conversion Corporation, "Enhancement Mode Power Transistor," EPC2036 datasheet, Apr. 2015 [Online] Available: http://epc-co.com/epc/documents/datasheets/EPC2036_datasheet.pdf

[10.6] Efficient Power Conversion Corporation, "Enhancement Mode Power Transistor," EPC2037 datasheet, Apr. 2015 [Online] Available: http://epc-co.com/epc/documents/datasheets/EPC2037_datasheet.pdf

[10.7] D. Reusch, J. Strydom, and A. Lidow, "Highly Efficient Gallium Nitride Transistors Designed for High Power Density and High Output Current DC-DC Converters," *IEEE International Power Electronics and Application Conference (PEAC)*, pp.456-461, 2014.

[10.8] A. Lidow, M. A. de Rooij, "Performance Evaluation of Enhancement-Mode GaN transistors in Class-D and Class-E Wireless Power Transfer Systems," *Bodo's Power Systems*, May 2014, pp. 56–60.

Experimental Verification Setup

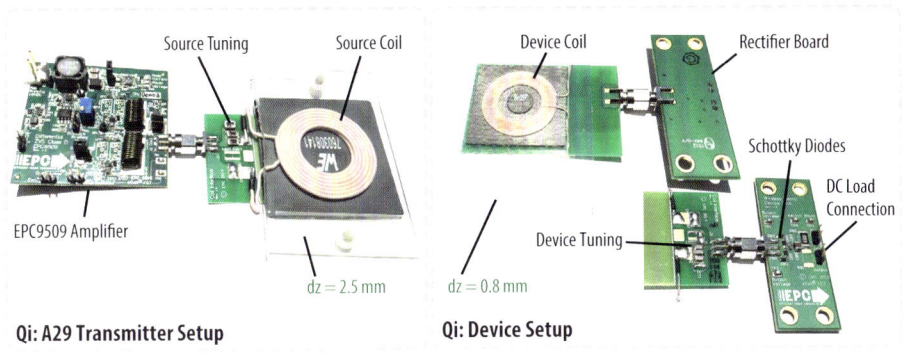

Qi: A29 Transmitter Setup   Qi: Device Setup

Experimental Verification Setup

eGaN FETs have already been experimentally validated to perform efficiently in A4WP based wireless power systems, so next they will be experimentally tested in a WPC Qi setup. Shown is the experimental Qi based system that comprises an A29 Qi compliant transmit coil [10.9] with tuning network (left), a Qi compliant device coil [10.10] connected to a tuning network, and full-bridge Schottky diode rectifier (right). The coils are mechanically connected to each other and the source coil connected to a modified EPC9509 [10.11] differential mode amplifier where the ZVS inductors have been disconnected and will be operated in bypass mode. The Qi prescribed spacing and alignment between the coils is set using Plexiglas spacers and nylon screws.

The supply voltage to the amplifier will be adjusted to yield the required $7\ V_{DC}$ on the device output as per the standard under test #20 for the receiver. The Qi A29 standard states that the source coil must be driven by a full bridge converter, hence the EPC9509 was chosen, and operated at a frequency of 130 kHz.

Similar to the A4WP system level tests, the experimental Qi setup will be tested and the temperature monitored to ensure the devices and/or gate drivers remain below 100°C. Per the Qi standard, the supply voltage to the amplifier also cannot exceed $12\ V_{DC}$.

[10.9] Würth Elektronik "WE-WPCC Wireless Power Charging Transmitter Coil," 760308141 datasheet [Online] Available:http://katalog.we-online.com/pbs/datasheet/760308141.pdf

[10.10] Würth Elektronik "WE-WPCC Wireless Power Charging Receiver Coil," 760308102210 datasheet [Online] Available: http://katalog.we-online.com/pbs/datasheet/760308102210.pdf

[10.11] [Online] Available: http://epc-co.com/epc/Products/DemoBoards/EPC9509.aspx

Experimental Results

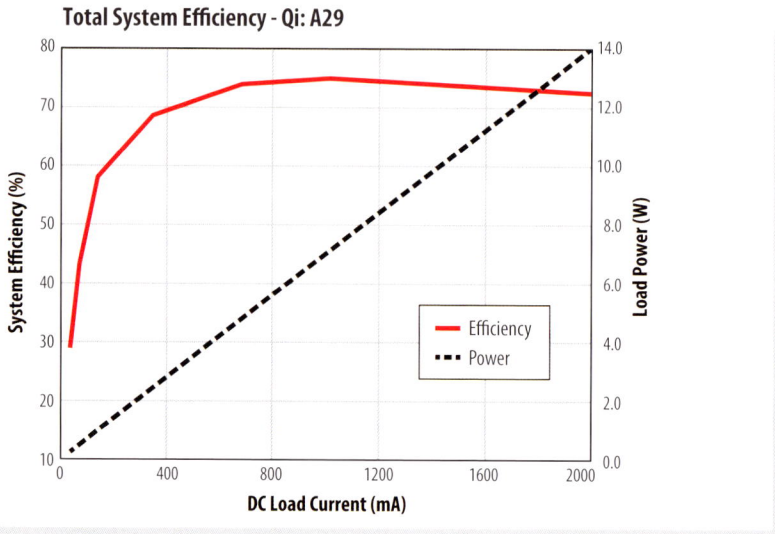

Total System Efficiency - Qi: A29

Experimental Results

The experimental result of the Qi setup is shown in the graph. The test started with a high device DC load resistance that was decreased in discrete steps down to approximately 3.5 Ω while the output voltage was maintained at 7 V. The black dashed line shows the output power trend for this test. The system efficiency, DC input power to the amplifier to the device DC output, including gate driver power is shown by the red trace. The peak efficiency is 75% at 1 A load (7 W). The efficiency drops off at higher current as the EPC2108 devices used in the EPC9509 need to have a higher current rating for this specific source coil to operate at the higher power, and are more suited to operate as a single-ended amplifier in this setup.

This experimental setup demonstrated that eGaN FETs are capable of operating a Qi based coil set despite not being correctly sized for this specific application. This was intentional to demonstrate that a single amplifier can ultimately be used to realize a multi-mode wireless power system. For a true multi-mode system, the coil and amplifier design will need to be closer matched by design.

Thermal Performance

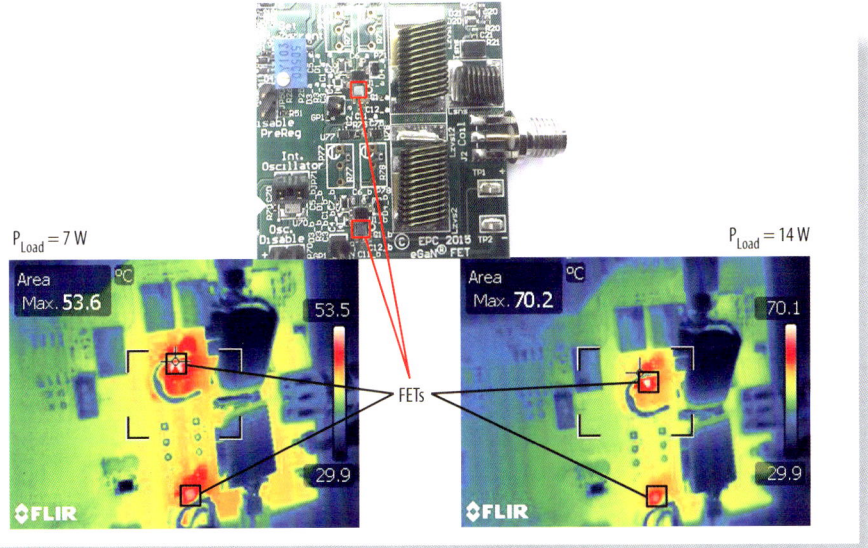

Thermal Performance

The thermal performance of the EPC9509 amplifier operating the Qi coil set is shown for two power levels of 7 W and 14 W. Again, despite the eGaN FETs not being correctly sized for this specific application, the devices operate well within target thermal specifications. In this case the peak current in the FETs at full power (14 W) was 1.7 A.

Multi-Mode Approach Summary

A multi-mode capable wireless power transfer system:

- Improves the user experience

- Increases the system cost – yet is lower than a muti-amplifier approach

- ZVS Class D supports both low and high frequency modes

- eGaN FETs used in the ZVS Class D amplifier will yield the highest efficiency at both low and high frequency

- A multi-mode coil structure must be designed specifically for the amplifier to yield highest performance for all standards

Multi-Mode Approach Summary

An analysis of the various wireless power standards and system architectures has suggested that a multi-mode approach is feasible and may improve the user's wireless power transfer experience. Unfortunately, a multi-mode approach will increase cost, but these costs can be contained with careful planning and design, and are lower than a multi-amplifier approach.

The ZVS class D amplifier topology is the only topology that supports both modes in a single amplifier and still offers high efficiency. The use of eGaN FETs in the amplifier will further ensure high efficiency even when the amplifier is operating at 315 kHz because of their ability to operate efficiently in high frequency hard-switching converters.

A specifically designed multi-mode coil structure is needed to ensure the highest efficiency for any of the wireless power modes and must be designed around the amplifier.

CHAPTER 11:

Control for Wireless Power Systems

The various building blocks for a wireless power transfer system have been discussed. Now those pieces need to be assembled into a functioning system. What will be covered is:

- System architecture
- Feedback measurement techniques
- Control parameter determination
- Control strategies

Wireless Power Control Overview

Up to this point the various building blocks of a wireless power system have been presented and discussed and include the amplifier, source coil, device coil, coil tuning, EMI filter, adaptive matching etc. The next step is to assemble the building blocks into a working wireless power system and provide an overview of methods to control the system. This discussion will focus on a ZVS class D amplifier based solution. What will not be covered is a detailed analysis of feedback control parameters for stable operation or an actual implementation, as these fall outside the scope of this work.

First the system architecture will be presented showing what the various building blocks are, and in what order they are assembled. This is needed to be able to draft applicable control strategies.

Techniques to measure various operating parameters of the system, such as coil current and coil voltage, will be presented. These are needed to accurately determine the tuned coil impedance that, in turn, determines the control method to be applied, such as constant coil current or power limit mode.

Having made various operating measurements, the means of converting those measurements into control parameters, such as coil impedance and power phase angle, will then be presented. Techniques for coil impedance determination using only the amplifier input power measurement provide accurate control parameters only if the amplifier has very high efficiency. It was demonstrated that in constant current mode and low power the amplifier efficiency can drop dramatically, causing significant errors in the control parameters that can lead to excessive coil currents, or even failure in extreme cases. New techniques for more accurate impedance determination are needed and will be presented.

Using the control parameters, two control strategies for the ZVS class D amplifier will be then presented.

Source System Block Diagram

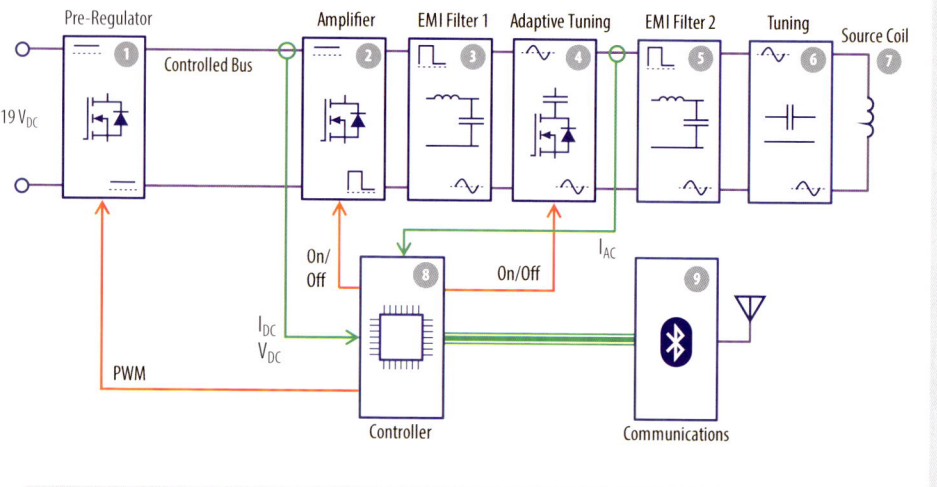

Source System Block Diagram

First we present a wireless power system source architecture suitable to realize an A4WP compliant product. There are nine major building blocks for this architecture:

1. Pre-regulator
2. Amplifier – such as a ZVS class D
3. EMI filter – for low order harmonics
4. Adaptive tuning
5. Second EMI filter – primarily for high frequency
6. Source coil tuning
7. Source coil
8. Controller – typically digital
9. Communications module – Bluetooth Low Energy (BLE)

Most of the building blocks have already been discussed, but there are a few new ones including a controller and Bluetooth communications module. The controller is typically some form of digital micro-controller or processor that oversees the entire system. It receives the various measurements from the system, processes them and provides control signals to the amplifier and pre-regulator. If an adaptive matching circuit is present, the controller will also provide the necessary control signals. Included in the control signals is information provided via the Bluetooth module from the device, or devices, either connected, or requesting to be connected. This information is also used to make adjustments in the control signals based on compliance with a standard, such as A4WP [11.1], and other parameters.

[11.1] *A4WP Wireless Power Transfer System Baseline System Specification (BSS)*, A4WP-S-0001 v1.3.1, February 25, 2015.

Device System Block Diagram

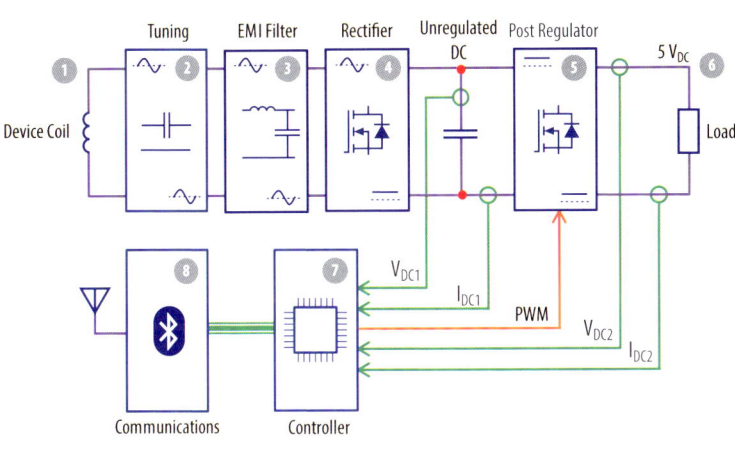

Device System Block Diagram

Next we present a wireless power system device architecture suitable to realize an A4WP compliant product. There are eight major building blocks for this architecture which in order are:

1. Device coil
2. Device coil tuning
3. EMI filter
4. Rectifier – typically a Schottky diode full bridge
5. Post-regulator – typically a Buck converter
6. Load – such as a cell phone battery or USB port
7. Controller – typically digital
8. Communications module – Bluetooth Low Energy (BLE)

Due to cost and size constraints, blocks 4, 5, 7 and 8 are typically monolithically integrated for power levels below 10 W. Above 10 W, the rectifier losses will become too high and discrete solutions become the norm.

Device System Block Diagram - *continued*

Up to this point the device discussion has ended at the un-regulated output of the rectifier with the voltage depending on coil conditions and load power demand. This un-regulated voltage must be converted into a stable voltage for use in most applications. Depending on the application, the fixed output voltage can be 5 V. A buck converter is used to provide voltage regulation as most usable operating conditions will lead to a high enough input voltage to reduce down to 5 V. Furthermore, some device applications have power dissipation limits imposed on them, such as 0.5 W for smart phones, thereby eliminating linear regulators as an option for the post regulator, and making the buck converter the best choice. The buck converter is also the simplest and most cost effective solution.

The controller receives the operating measurements and provides the control signal for the buck converter. It also communicates with the rest of the device and with the source via the BLE. This is necessary to connect to the source and inform the source of power requirements.

High Frequency Current Measurement

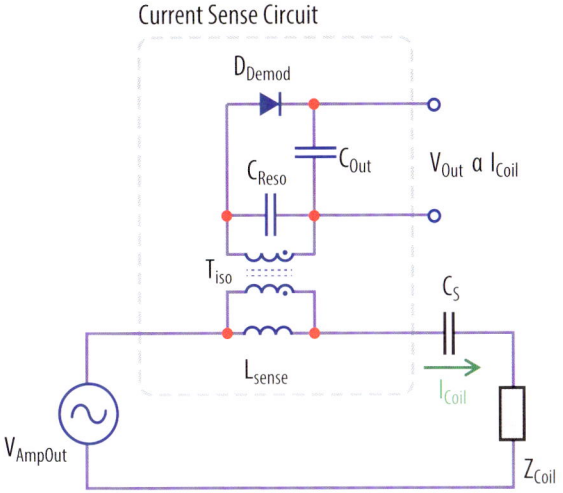

Current Sense Circuit

High Frequency Current Measurement

The controller requires measurements of various operating conditions in the wireless power system, and the most important is the coil current (I_{coil}). This high frequency current is not trivial to measure and classic shunt techniques result in an output voltage that is very small and difficult to process. Specialized isolated current sensors can overcome the low output voltage issue, but become prohibitively expensive. A simpler technique is needed to measure the coil current.

Shown in the figure is a schematic for the measurement of the high frequency coil current. The output of this circuit is a voltage that is proportional to the magnitude of the current. The circuit features full galvanic isolation so it can be used in single-ended or differential-mode amplifier topologies. The measurement circuit uses a simple diode rectifier envelope tracker typically found in low cost amplitude modulation (AM) radios. The key feature of this measurement circuit is the large signal obtained that makes it easy to provide to a digital controller with an analog input.

The circuit works by using an inductor (L_{sense}) as a high frequency current sensor. The inductance value of the sensor can be very small at between 20 nH and 100 nH for 6.78 MHz systems and depends on the magnitude of the current being measured. A micro-transformer (T_{iso}) has its primary winding connected across the current sensor and a capacitor (C_{Reso}) connected across the secondary. This capacitor's value is chosen so that it resonates with the leakage inductance of the transformer at 6.78 MHz, converting the reflected current into a large voltage which can now be rectified using a single Schottky diode (D_{Demod}). A Schottky diode is used primarily for its low voltage drop and zero reverse recovery, thereby improving low current measurement performance.

High Frequency Current Measurement - *continued*

It is also important to resonate with the leakage inductance of the transformer, and not the self or magnetizing inductance, as doing so will reflect back across the transformer and increase the impedance of the current sensor. This will effectively disconnect the coil from the amplifier. A capacitor (C_{out}) is used to smooth out the rectified voltage that can now be provided directly to the controller for processing.

This circuit can be made physically very small in relation to the balance of the amplifier, and can be realized using standard components. It also can be calibrated using firmware within the controller, making it easier to manufacture. The added inductance in the series path to the coil can easily be compensated for by re-calculating the series tuning capacitor (C_s), and in the worst case approximately $5j \, \Omega$ will be added to the tuned coil circuit impedance. The small value of the current sense inductor further ensures that the ESR will be low, and hence power dissipation will also be low.

Measured HF Current Sense Performance

Current Sensor Transfer Function

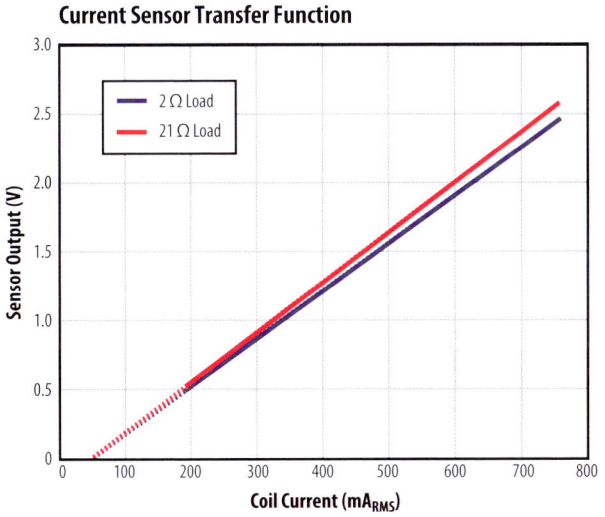

Measured HF Current Sense Performance

To verify the performance of the current sense, a practical circuit was constructed using the following component values:

L_{sense} = 110 nH (CoilCraft part # 2222SQ-111JE) [11.2]

T_{iso} = 10 μH 1:1 coupled chip inductor (CoilCraft part # PFD3215-103ME) [11.3]

C_{Reso} = 680 pF ceramic

D_{Demod} = 40 V, 30 mA Schottky diode (Diodes Inc. part # SDM03U40) [11.4]

C_{out} = 1nF ceramic

The circuit was loaded with a 10 kΩ resistor and tested using an emulating load comprising an inductor, series capacitor, and series load resistor similar to the one used for amplifier testing. The graph shows the results of the experiment for two different AC load resistance settings spanning a range between 200 mA and 750 mA. The dotted line extrapolates to the minimum current that can be measured using this setup due to the voltage drop of the rectifier diode.

[11.2] "Square Air Core Inductors," CoilCraft datasheet PFD3215, Document 720 -1, Revised 16 August 2012, [Online] Available: http://www.coilcraft.com/pdfs/1515sq.pdf

[11.3] "Coupled Chip Inductors," CoilCraft datasheet PFD3215, Document 982F-1, Revised 5 December 2012, [Online] Available: http://www.coilcraft.com/pdfs/pfd3215_flyback.pdf

[11.4] "Surface Mount Schottky Barrier Diode," Diodes Inc. datasheet PFD3215, Document number: DS30392 Rev. 9 - 2, Revised January 2009, [Online] Available: http://www.diodes.com/_files/datasheets/ds30392.pdf

Coil Voltage, Current, and Phase Detect

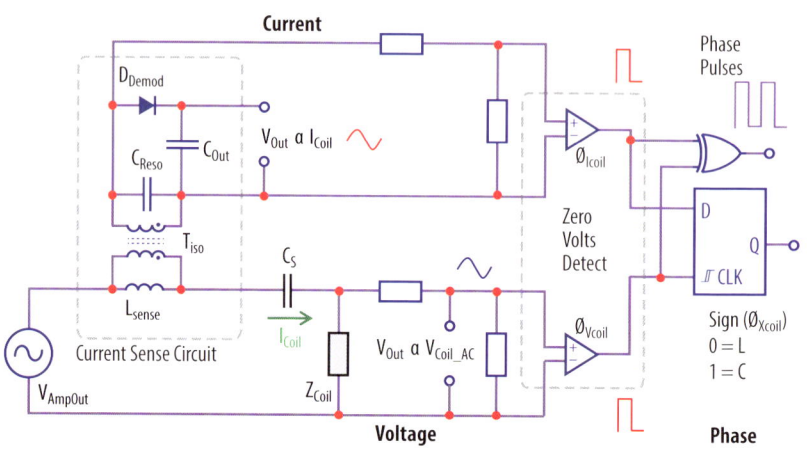

Coil Voltage, Current, and Phase Detect

To be able to determine the coil impedance various measurements are needed that include, coil current, coil voltage, and the phase angle between the coil voltage and current. Measurement of the coil current magnitude has been presented. Using a simple high resistance value divider network and a diode rectifier, the coil voltage magnitude can be determined. For a differential-mode setup the coil voltage will also need to be measured differentially and processed using a high bandwidth operational amplifier for each phase, or it can be fed directly into two separate analog inputs on the controller to be processed internally. Any phase shift between the two measured voltage signals will result in a constant offset that can be calibrated out.

The determination of the phase angle between the coil voltage and current is a two-step process. The high frequency scaled voltages (V_{OutV} and V_{OutI}) of the respective measurements are each fed into a comparator that acts as a zero crossing detect. It is important to use similar voltage scaling for each comparator to minimize phase offset error. These two digital zero crossing detect signals are then fed into a D-flip-flop. For the setup shown, the output of the D-flip-flop will become high if the current leads the voltage (i.e. capacitive), and will become low if the current lags the voltage (i.e. inductive), and is used to determine the sign of the phase angle. The magnitude of the phase angle can be derived by inputting the two digital zero crossing detect signals into an exclusive OR logic gate (XOR). The output of the XOR logic gate will be a series of pulses that can be passed through a low-pass filter to yield a voltage that is proportional to the phase angle magnitude.

Determination of Coil Impedance Overview

Measurements:

- I_{coil} – RMS coil current
- V_{coil} – RMS coil voltage
- V_{AMP} – Amplifier supply voltage
- I_{AMP} – Amplifier supply current
- $\emptyset_{Icoil}$ – Coil current phase

Assumptions:

Power Amplifier has high efficiency when operating at full power:
$P_{AMP} \approx P_{coil}$, $\eta_{AMP} > 95\%$

Processing:

1. $V_{coil} = (\sqrt{2/\pi}) \cdot V_{AMP}$ (2x for differential mode)
2. $P_{coil} = V_{coil\_RMS} \cdot I_{coil\_RMS} \cdot \cos(\emptyset_{Icoil})$ OR $= V_{AMP} \cdot I_{AMP}$
3. $R_{coil} = P_{coil}/(I_{coil})^2$
4. $|Z_{coil}| = V_{coil}/ I_{coil}$
5. $|X_{coil}| = \sqrt{(|Z_{coil}|^2 - R_{coil}^2)}$
6. X sign= Data latch $\emptyset_{Icoil}$ to oscillator

Determination of Coil Impedance Overview

With the ability to measure the key coil parameters, those parameters need to be converted into control parameters. In addition to the measured coil parameters, the DC input voltage (V_{AMP}) and current (I_{AMP}) to the amplifier also need to be measured. These are needed to verify some of the more difficult to calculate parameters, and to ensure the amplifier is operating within its design limits.

A quick method to determine the coil power is to calculate it from the power supplied to the amplifier. This only works if the amplifier has high efficiency (> 95%) and, based on the experimental results presented earlier, this is not always the case.

Determination of Coil Impedance Overview - *continued*

If the system is operating with the coil on-resonance, then the coil voltage (V_{coil}) can also be determined from the supply voltage (V_{AMP}). The equation for a ZVS class D amplifier is given in the example shown. The power (P_{coil}) delivered into the coil can also be determined from the measured coil voltage (V_{coil}) and current (I_{coil}) magnitudes multiplied together, including the cosine of the phase angle ($\cos(\varnothing_{coil})$). The quickest method to determine the cosine of the phase angle in a controller is to use a lookup table. The coil determined power (P_{coil}), and the amplifier measured power (P_{AMP}), can also be used to determine the efficiency of the amplifier. This information can be used to adjust the operating point to maximize efficiency.

The resistance (R_{coil}) of the coil can now be calculated using the equation shown, and will work only if the current is of sufficient magnitude to avoid dividing by zero. This should not be a problem as there will always be a current in the coil of sufficient magnitude under most and normal operating conditions.

The magnitude of the coil impedance ($|Z_{coil}|$) can be calculated from the coil voltage and current. This can then be used to determine the magnitude of the reactive component ($|X_{coil}|$) of the coil impedance. The magnitude of the reactive component of the coil impedance, together with the power phase angle sign (X_{sign}), can be used to determine adaptive matching control parameters (e.g. which capacitors to switch in or out to retune the coil).

For a wireless power system only the coil current, coil power, and amplifier supply voltage are needed to sufficiently control the amplifier. All other parameters are needed for the system to make decisions and do not need to be in the critical control path.

Control Architecture – Amplifier Supply Voltage Regulation

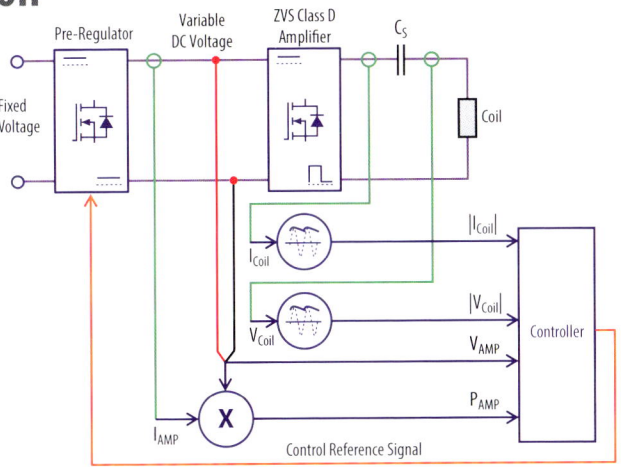

Control Architecture – Amplifier Supply Voltage Regulation

The first method that can be used to control a ZVS class D amplifier is presented here. This control method works by adjusting the supply voltage to the amplifier which in turn controls either the coil current or coil power, depending on the coil impedance and mode of operation. This control method is suitable for both single-ended and differential-mode amplifiers and has been implemented on both the EPC9509 [11.5] and EPC9510 [11.6] demonstrations boards.

For this control method only three measurements are needed: the magnitude of the coil current ($|I_{coil}|$), the amplifier DC supply voltage (V_{AMP}), and the amplifier DC supply current (I_{AMP}). Using a multiplier, the amplifier power (P_{AMP}) can be determined.

A combiner is used to determine in which mode the amplifier must operate. This can only be one of three: constant coil current, constant power, or constant voltage.

The constant voltage mode has the highest priority and is used to limit the supply voltage to the amplifier. This can occur if the magnitude of the coil impedance is higher than the amplifier operating limit. An example would be a class 3 coil presenting a coil impedance of 45-100j Ω to the EPC9509 amplifier, which would need a voltage that far exceeds the maximum of 52 V to drive 16 W into the coil. This mode of operation does not comply with the A4WP standard.

Control Architecture – Amplifier Supply Voltage Regulation - *continued*

The constant power mode has the next priority for control. From the experimental data it was seen that since the amplifier has high efficiency when operating in the constant power region, the amplifier power measurement can be used as a control parameter for coil power. Since power control is not critical, the A4WP standard allows for higher power to be delivered, and therefore any variation due to amplifier efficiency is acceptable.

The constant current mode has the lowest priority, but the highest control bandwidth requirement. This is due to small changes in voltage that can lead to large current swings in the coil. In this mode power delivered to the load becomes reduced as the coil becomes under-utilized.

The combiner determines which control mode to operate in, and provides a reference control signal to the pre-regulator. The pre-regulator can be a SEPIC converter, or any suitable converter that can control the amplifier voltage.

[11.5] [Online] Available: http://epc-co.com/epc/Products/DemoBoards/EPC9509.aspx

[11 6] [Online] Available: http://epc-co.com/epc/Products/DemoBoards/EPC9510.aspx

Differential-Mode ZVS Class D Phase Shift Control

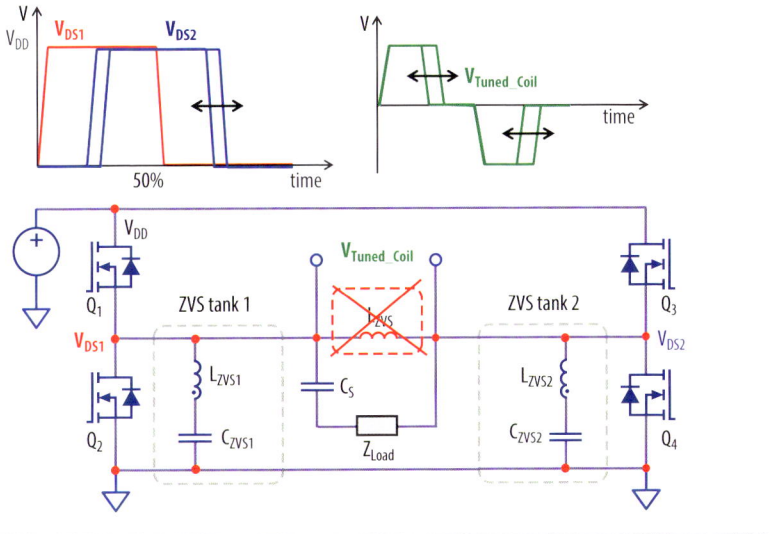

Differential-Mode ZVS Class D Phase Shift Control

Another method that can be used to control the ZVS class D amplifier is phase shift control. This method will only work for the differential-mode amplifier that must also be configured as two individual single-ended amplifiers due to each single-end amplifier becoming separately controlled.

For this method each amplifier operates at a fixed frequency, 6.78 MHz in this case, and with a fixed duty-cycle of 50%, as has been the case for all of the ZVS class D amplifiers discussed. The control signal is then phase modulated between the two single-ended amplifiers, as shown in the top left where the red signal is the reference for the left single-ended amplifier, and the blue is the phase shifted signal for the second single-ended amplifier. The result of phase shifting the signals is shown in green in the top right, yielding an equivalent output signal with a varying duty-cycle that is proportional to the phase difference between the two amplifiers. Maintaining 50% duty cycle and a fixed frequency is critical for ensuring proper operation of each of the amplifiers.

This control method has the advantage over the amplifier supply regulation method because it can react faster to a change in the control signal input, thus making it easier to control. This reduces the control bandwidth requirements of the controller. The disadvantage of this control method is its potential to increase EMI in the lower frequency spectrum, including spreading it out so frequencies can appear that are not present in the amplifier supply voltage control method.

Control Architecture – Phase Shift Control

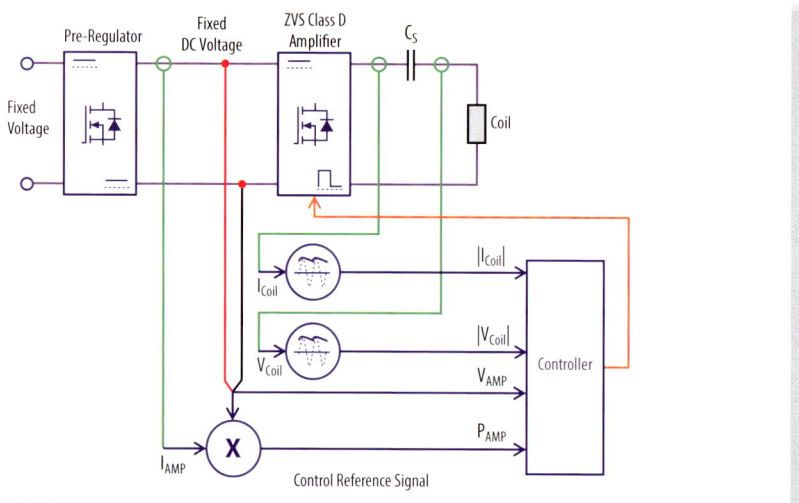

Control Architecture – Phase Shift Control

Shown is the implementation of the phase shift controller that is based on the amplifier supply voltage controller. Similar parameters are measured, but instead of controlling the pre-regulator this method directly controls the ZVS class D amplifiers. This simplifies the pre-regulator allowing a simple boost converter to be used instead. The output of the combiner from the amplifier supply voltage controller is used as a reference to generate the phase shifted signals. This can be single-edge where only the second amplifier phase is shifted, or dual-edge driven where both amplifiers are phase shifted relative to each other. The latter method is preferred as it yields a symmetrical impact to both amplifiers.

Implementation strategy

Most of the wireless power transfer system control can be implemented using a relatively low cost, low power micro-controller. Functions to be included are:

- Coil operating parameter determination
- PWM generation for the pre-regulator
- Amplifier on/off
- Analog signal processing (low frequency)
- Adaptive matching

Implementation Strategy

In summary, various control strategies have been presented. Most of the control functions can be implemented in a micro-controller, thereby keeping costs down for the system. These functions include coil parameters determination such as impedance and phase angle, PWM generation for the pre-regulator so an external controller is not needed, amplifier on/off control for A4WP compliance to detect devices, and various analog signal processing for the system to make adaptive matching decisions.

CHAPTER 12:

Summary

The wireless power transfer convenience factor is addressed by the highly resonant, loosely coupled approach of the A4WP standard

Both the class E and ZVS class D amplifiers are suitable for wireless power transfer at 6.78 MHz, but ZVS class D is superior

The ZVS class D amplifier using eGaN FETs reduces wireless power cost

eGaN FETs and integrated circuits contribute to superior amplifier performance and are enabling efficient, low cost wireless power

Summary

The concept of wireless power transfer that uses magnetic fields as a medium has been introduced. Magnetic fields were chosen as they offer ease of use and are considered safe. Potential applications, such as mobile computing and medical that could be significantly impacted by this technology, were presented.

A brief overview of the various wireless power transfer standards were introduced leading up to the highly resonant, loosely coupled approach, which was deemed the most suitable for further investigation. This led to the introduction of the highly resonant coil sets that form the core of the wireless power system, and how these coils are typically tuned.

Next, various amplifier topologies suitable for use in wireless power transfer systems, including class E, traditional voltage-mode class D, current-mode class D, and ZVS class D topologies were introduced. These amplifiers all operate as switched-mode converters to ensure the highest possible efficiency. The design of these amplifiers was covered with respect to the needs of wireless power transfer.

Summary - *continued*

Suitable devices for the amplifiers were presented, and a means to quickly compare the potential performance benefits using a figure of merit approach was discussed. Then, experimental versions of the amplifiers operating on-resonance were presented. From those experimental results, it was clear that the ZVS class D topology, fitted with eGaN FETs, exhibited superior performance over the other amplifiers.

Following on-resonance experimental verification, the elements that contribute to the convenience factor of wireless power transfer were discussed. The potential end-use scenarios showed that multiple simultaneous loads and the introduction of foreign metal objects affect the coil impedance. To address these end-use conditions, the A4WP Rezence standard prescribes a wide operating impedance range for the amplifier.

The Rezence standard was used as a guide to evaluate the capability of the amplifiers. Testing showed that eGaN FETs consistently outperformed MOSFETs, regardless of the amplifier tested, and that the ZVS class D amplifier was capable of operating over a significantly wider impedance range than the class E amplifier. System-level tests were also conducted using actual coils to deliver power. Two types of A4WP compliant coils were tested; a class 2 and a class 3. The results show good efficiency for the class 2 system and the ability of eGaN FETs used in a ZVS class D amplifier to realize a fully class 2 compliant system without the need for adaptive matching.

A method was developed to further improve the efficiency of the ZVS class D amplifier by eliminating the reverse recovery losses of the gate driver bootstrap power supply. This is accomplished by replacing the diode with an eGaN FET that is switched synchronously with the lower device. Experimental testing yielded dramatic improvements at both 6.78 MHz and 13.56 MHz, with the operating temperature dropping by as much as 11°C and 31°C respectively.

A radiated EMI analysis of a wireless power systems showed that the class E amplifier is susceptible to generating high levels of even-order harmonics that are difficult to remove, whereas the ZVS class D is not susceptible to generating these harmonics. The process of designing an EMI filter that does not impact the coil was also presented with an analysis of its performance capability. Modification of the filter to account for common-mode EMI was then given.

The multitude of wireless standards has also led to consumer confusion and, to address this issue, a multi-mode approach to wireless power transfer was presented. It was shown that it is possible to design a cost-effective multi-mode source unit that can drive source coils of the various standards using a single amplifier.

Finally, various control strategies for the ZVS class D amplifier were presented together with methods to measure various control parameters such as coil current and power phase angle. The implemented control strategy used on the demonstration boards was also given including a new phase shift control method.

Now we can finally cut the cord and go wireless!

APPENDIX A

Efficient Power Conversion (EPC) Product Selector Guide

For all devices, Max $V_{GS} = 6$ V
For all devices $Q_{RR} = 0$ nC

eGaN® FETs

Part Number	Configuration	V_{DS}	Max $R_{DS(on)}$ (mΩ) @5V_{GS}	Q_G typ (nC)	Q_{GS} typ (nC)	Q_{GD} typ (nC)	Pulsed I_D (A)	LGA Package (mm)
EPC2100	Dual Asymmetric	30	8 2	3.5 15	1.4 4.6	0.57 2.6	100 400	6.1 x 2.3
EPC2023	Single	30	1.3	20	5.8	1.9	590	6.1 x 2.3
EPC2024	Single	40	1.5	19	6.4	2.0	550	6.1 x 2.3
EPC2030	Single	40	2.4	18	5.2	3.4	495	4.6 x 2.6
EPC2015C	Single	40	4	8.7	3.0	1.4	235	4.1 x 1.6
EPC2015	Single	40	4	10.5	3.0	2.2	150	4.1 x 1.6
EPC2014C	Single	40	16	2.0	0.70	0.30	60	1.7 x 1.1
EPC2014	Single	40	16	2.5	0.67	0.48	40	1.7 x 1.1
EPC8004	Single	40	110	0.370	0.120	0.047	7.5	2.1 x 0.85
EPC8007	Single	40	160	0.302	0.097	0.025	6	2.1 x 0.85
EPC8008	Single	40	325	0.177	0.067	0.012	2.9	2.1 x 0.85
EPC2020	Single	60	2	16	5.0	2.0	470	6.1 x 2.3
EPC2031	Single	60	2.6	17	5.2	3.2	450	4.6 x 2.6
EPC2035	Single	60	45	0.88	0.3	0.2	24	0.9 x 0.9
EPC2101	Dual Asymmetric	60	11.5 2.7	2.7 12	1 3.7	0.50 2.5	80 350	6.1 x 2.3
EPC2102	Dual	60	4.4	6.8	2.3	1.4	215	6.1 x 2.3
EPC2108	Dual with Bootstrap	60	190	0.22	0.085	0.045	5.5	1.35 x 1.35
EPC8009	Single	65	130	0.370	0.120	0.055	7.5	2.1 x 0.85
EPC8005	Single	65	275	0.218	0.077	0.018	3.8	2.1 x 0.85
EPC8002	Single	65	530	0.141	0.059	0.009	2	2.1 x 0.85
EPC2021	Single	80	2.5	15	3.8	2.1	420	6.1 x 2.3

For all devices, Max $V_{GS} = 6$ V
For all devices $Q_{RR} = 0$ nC

eGaN® FETs (continued)

Part Number	Configuration	V_{DS}	Max $R_{DS(on)}$ (mΩ) @5V_{GS}	Q_G typ (nC)	Q_{GS} typ (nC)	Q_{GD} typ (nC)	Pulsed I_D (A)	LGA Package (mm)
EPC2029	Single	80	3.2	13	4.0	2.5	360	4.6 x 2.6
EPC2105	Dual Asymmetric	80	14.5 / 3.5	2.5 / 10	1 / 3.2	0.50 / 2	75 / 320	6.1 x 2.3
EPC2103	Dual	80	5.5	6.5	2.0	1.3	195	6.1 x 2.3
EPC2039	Single	80	6	22	2	0.63	50	1.35 x 1.35
EPC2104	Dual	100	6.3	7	2.0	1.2	165	6.1 x 2.3
EPC2022	Single	100	3.2	13	3.7	2.0	360	6.1 x 2.3
EPC2032	Single	100	4	14	4.2	3.1	340	4.6 x 2.6
EPC2001C	Single	100	7	7.5	2.4	1.2	150	4.1 x 1.6
EPC2001	Single	100	7	8.0	2.3	2.2	100	4.1 x 1.6
EPC2016C	Single	100	16	3.4	1.1	0.55	75	2.1 x 1.6
EPC2016	Single	100	16	3.8	0.99	0.7	50	2.1 x 1.6
EPC2007C	Single	100	30	1.6	0.6	0.3	40	1.7 x 1.1
EPC2007	Single	100	30	2.1	0.52	0.61	25	1.7 x 1.1
EPC2036	Single	100	65	0.7	0.17	0.14	18	0.9 x 0.9
EPC2106	Dual	100	6	0.73 / 0.76	0.22 / 0.24	0.165	18	1.35 x 1.35
EPC8010	Single	100	160	0.360	0.130	0.060	7.5	2.1 x 0.85
EPC8003	Single	100	300	0.315	0.110	0.034	5	2.1 x 0.85
EPC2107	Dual with Bootstrap	100	320	0.16	0.065	0.04	3.8	1.35 x 1.35
EPC2037	Single	100	6	550	0.115	0.030	2.4	EPC9051
EPC2038	Single	100	6	2800	0.044	0.016	0.5	N/A
EPC2110	Dual, Common Source	120	6	60	0.8	0.25	20	N/A
EPC2033	Single	150	7	10	3.5	1.7	260	4.6 x 2.6
EPC2018	Single	150	25	5.0	1.3	1.7	60	3.6 x 1.6
EPC2034	Single	200	10	8.5	2.6	1.4	140	4.6 x 2.6
EPC2010C	Single	200	25	3.7	1.3	0.7	90	3.6 x 1.6
EPC2010	Single	200	25	5.0	1.3	1.7	60	3.6 x 1.6
EPC2019	Single	200	50	1.8	0.60	0.35	42	2.7 x 0.95
EPC2012C	Single	200	100	1.0	0.30	0.20	22	1.7 x 0.9
EPC2012	Single	200	100	1.5	0.33	0.57	15	1.7 x 0.9
EPC2025	Single	300	150	1.9	0.61	0.30	20	1.95 x 1.95
EPC2027	Single	450	400	1.7	0.60	0.25	12	1.95 x 1.95

DrGaN^{PLUS}

Part Number	Description	V$_{DS}$ (max)	I$_D$ (max RMS)	Featured Product
EPC9201	PCB-based half-bridge circuit module	30	20	EPC2015/EPC2023
EPC9203	PCB-based half-bridge circuit module	80	40	EPC2021

Half Bridge Development Boards

Part Number	Description	V$_{DS}$ (max)	I$_D$ (max RMS)	Featured Product
EPC9036	Power Stage evaluation of monolithic GaN half bridge	30	25	EPC2100
EPC9018	Half Bridge Plus Driver for low duty cycle applications	30	35	EPC2015/EPC2023
EPC9031	Half Bridge Plus Driver	30	40	EPC2023
EPC9005C	Half Bridge Plus Driver	40	7	EPC2014C
EPC9005	Half Bridge Plus Driver	40	7	EPC2014
EPC9001	Half Bridge Plus Driver	40	15	EPC2015
EPC9016	Half Bridge Plus Driver for low duty cycle applications	40	25	EPC2015
EPC9032	Half Bridge Plus Driver	40	35	EPC2024
EPC9049	Half Bridge Plus Driver	60	4	EPC2035
EPC9038	Power Stage evaluation of monolithic GaN half bridge	60	20	EPC2102
EPC9037	Power Stage evaluation of monolithic GaN half bridge	60	22	EPC2101
EPC9033	Half Bridge Plus Driver	60	30	EPC2020
EPC9057	Half Bridge with Gate Drive	80	6	EPC2039
EPC9039	Power Stage evaluation of monolithic GaN half bridge	80	17	EPC2103
EPC9041	Power Stage evaluation of monolithic GaN half bridge	80	20	EPC2105
EPC9019	Half Bridge Plus Driver for low duty cycle applications	80	20	EPC2001/EPC2021
EPC9046	Half Bridge Plus Driver	80	22	EPC2029
EPC9034	Half Bridge Plus Driver	80	27	EPC2021
EPC9050	Half Bridge Plus Driver	100	2.5	EPC2036
EPC9055	Half Bridge Plus Driver	100	3	EPC2106
EPC9006	Half Bridge Plus Driver	100	5	EPC2007

Half Bridge Development Boards (continued)

Part Number	Description	V$_{DS}$ (max)	I$_D$ (max RMS)	Featured Product
EPC9010C	Half Bridge Plus Driver	100	7	EPC2016C
EPC9010	Half Bridge Plus Driver	100	7	EPC2016
EPC9002C	Half Bridge Plus Driver	100	10	EPC2001C
EPC9002	Half Bridge Plus Driver	100	10	EPC2001
EPC9040	Power Stage evaluation of monolithic GaN half bridge	100	15	EPC2104
EPC9017	Half Bridge Plus Driver for low duty cycle applications	100	20	EPC2001
EPC9035	Half Bridge Plus Driver	100	25	EPC2022
EPC9013	Multiple Half Bridges in Parallel	100	35	EPC2001
EPC9047	Half Bridge Plus Driver	150	12	EPC2033
EPC9004C	Half Bridge Plus Driver	200	3	EPC2012C
EPC9004	Half Bridge Plus Driver	200	3	EPC2012
EPC9014	Half Bridge Plus Driver	200	4	EPC2019
EPC9014	Half Bridge Plus Driver	200	4	EPC2019
EPC9003C	Half Bridge Plus Driver	200	5	EPC2010C
EPC9003	Half Bridge Plus Driver	200	5	EPC2010
EPC9042	Half Bridge Plus Driver	300	3	EPC2025
EPC9044	Half Bridge Plus Driver	400	1.5	EPC2027

High-frequency Half Bridge Development Boards

Part Number	Description	V$_{DS}$ (max)	I$_D$ (max RMS)	Featured Product
EPC9024	Half Bridge Plus Driver	40	4.4	EPC8004
EPC9027	Half Bridge Plus Driver	40	3.5	EPC8007
EPC9028	Half Bridge Plus Driver	40	2.2	EPC8008
EPC9022	Half Bridge Plus Driver	65	1.6	EPC8002
EPC9025	Half Bridge Plus Driver	65	2.2	EPC8005
EPC9029	Half Bridge Plus Driver	65	3.5	EPC8009
EPC9023	Half Bridge Plus Driver	100	2.2	EPC8003
EPC9030	Half Bridge Plus Driver	100	3.2	EPC8010

Demonstration Boards

Part Number	Description	V$_{IN}$	V$_{OUT}$	I$_{OUT}$	Featured Product
EPC9051	High Frequency Class-E Power Amplifier	0 V - 40 V	—	1 A	EPC2037
EPC9101	19 V to 1.2 V 1MHz Buck Converter	8 V - 19 V	1.2 V	18 A	EPC2015/EPC2014
EPC9102	48 V to 12 V 1/8th Brick Converter	36 V - 60 V	12 V	17 A	EPC2001
EPC9105	48 V to 12 V 1.2 MHz Intermediate Bus Converter	36 V - 60 V	12 V	30 A	EPC2001/EPC2015
EPC9106	150 W / 8 Ω Class D Audio Amplifier Demo System	—	—	—	EPC2016
EPC9107	28 V to 3.3 V Buck Converter	9 V - 28 V	3.3 V	15 A	EPC2015
EPC9111	A4WP compatible, ZVS Class-D Wireless Power System	8 V - 32 V	V$_{IN}$	10 A	EPC2014
EPC9112	A4WP compatible, ZVS Class-D Wireless Power System	8 V - 32 V	V$_{IN}$	6 A	EPC2007C/EPC2038
EPC9115	48 V to 12 V 1/8th Brick Converter	48 V - 60 V	12 V	42 A	EPC2020/EPC2021
EPC9118	48 V to 5 V 400 kHz Buck Converter	30 V - 60 V	5 V	20 A	EPC2001/EPC2021
EPC9506	Amplifier board for ZVS Class-D wireless system	8 V - 32 V	V$_{IN}$	10 A	EPC2014
EPC9507	Amplifier board for ZVS Class-D wireless system	8 V - 32 V	V$_{IN}$	6 A	EPC2007C/EPC2038
EPC9508	Amplifer board for ZVS Class-D wireless system	7 V - 36 V	V$_{IN}$	3 A	EPC8009/EPC2007
EPC9113	A4WP compatible, ZVS Class-D Wireless Power System	9 V - 24 V	52 V	1 A	EPC2108/EPC2036
EPC9114	A4WP compatible, ZVS Class-D Wireless Power System	9 V - 24 V	66 V	0.8 A	EPC2107/EPC2036
EPC9509	Amplifier board for ZVS Class-D wireless system	9 V - 24 V	52 V	1 A	EPC2108/EPC2036
EPC9510	Amplifier board for ZVS Class-D wireless system	9 V - 24 V	66 V	0.8 A	EPC2107/EPC2036

About the Author

Dr. Michael A. de Rooij is the Vice President of Applications Engineering at Efficient Power Conversion Corporation (EPC) headquartered in El Segundo, CA. Prior to joining EPC, he worked at Windspire Energy Inc., in the capacity of Director of Product Development where he helped develop the next generation of small vertical-axis wind turbine inverters. From 2002 to 2008, Dr. de Rooij worked as a Senior Engineer at the GE Global Research Center, located in Niskayuna, NY. Before joining GE, Dr. de Rooij conducted Post-Doctoral studies on power electronic integration, packaging and high-frequecy MOSFET gate driver at the Center for Power Electronic Systems (CPES) at the Virginia Polytechnic and State University. From 1998 through 2000, he was employed in the capacity of senior designer at IMV Victron in Groningen, The Netherlands, where his duties included the design of single-phase uninterruptible power supplies in the power range of 600 VA through 10 kVA.

Dr. de Rooij received his Ph.D. from the former Rand Afrikaans University (now known as the University of Johannesburg), South Africa in 1998. He is a Senior Member of the IEEE and is the author and co-author of over 25 international publications. Dr. de Rooij most recently co-authored *GaN Transistors for Efficient Power Conversion*, a textbook on gallium nitride technology. In addition, he has been granted 21 US and International patents and 19 US and International pending patent applications.

Dr. de Rooij's research interests and activities include wireless power transfer, solid-state high-frequency power converters and devices, uninterruptible power supplies, integration of power electronic converters, power electronic packaging, induction heating, photovoltaic converters, Magnetic Resonance Imaging (MRI) Systems, RF amplifiers and gate drivers with protection features.

About Efficient Power Conversion Corporation (EPC)

Efficient Power Conversion is the leader in enhancement-mode gallium nitride based power management devices. EPC was the first to introduce enhancement-mode gallium-nitride-on-silicon (eGaN) FETs as power MOSFET replacements in applications such as wireless power transfer, DC-DC converters, envelope tracking, RF transmission, power inverters, remote sensing technology (LiDAR), and class D audio amplifiers with device performance many times greater than the best silicon power MOSFETs.

eGaN® is a registered trademark of Efficient Power Conversion Corporation.

Index

A

B

C

T

V

W

Z